SpringerBriefs in Physics

More information about this series at http://www.springer.com/series/8902

Masa-aki Hashimoto • Riou Nakamura
E. P. Berni Ann Thushari • Kenzo Arai

Big-Bang Nucleosynthesis

Thermonuclear History in the Early Universe

Springer

Masa-aki Hashimoto
Department of Physics
Kyushu University
Fukuoka, Japan

E. P. Berni Ann Thushari
Department of Physics
Kyushu University
Fukuoka, Japan

Riou Nakamura
Kurume Institute of Technology
Fukuoka, Japan

Kenzo Arai
Department of Physics
Kumamoto University
Kumamoto, Japan

ISSN 2191-5423 ISSN 2191-5431 (electronic)
SpringerBriefs in Physics
ISBN 978-981-13-2934-0 ISBN 978-981-13-2935-7 (eBook)
https://doi.org/10.1007/978-981-13-2935-7

Library of Congress Control Number: 2018959104

This Springer imprint is published by the registered company Springer Nature Singapore Pte Ltd.
The registered company address is: 152 Beach Road, #21-01/04 Gateway East, Singapore 189721, Singapore

Preface

The origin of elements has been one of the most interesting issues in science, because it is closely connected with that of ourselves and the solar system. At present, this subject is considered to be related to the evolution of the universe. The Big Bang model, based on the theory of General Relativity, has proven to be the most reliable theory to describe the evolution of the universe. This is supported by the following observational facts: the expansion of the universe, the cosmic microwave background (CMB) radiation with a tiny fluctuation in temperature, and the primordial abundance of light elements, ^{4}He, D, and ^{7}Li.

Among many kinds of books concerning the Big Bang theory, we present a book which includes a detailed description of Big Bang nucleosynthesis (BBN) whose theme has not been explained well so far. BBN is the interdisciplinary field which incorporates nuclear physics, elementary particle physics, and astronomy. In particular, it is indispensable to obtain important information on the earth, such as the results of experiments by particle accelerators and observational findings from stars and galaxies. Originally, nucleosynthesis in the early universe had been studied fully by nuclear (particle) physicists, which were based on experimental data of nuclear fusion and β-decays. The basic idea that the light elements could be produced during the BBN epoch – *the first 3 min* – still holds now.

We are now in an amazing stage of development in modern cosmology. To derive matter density of the universe, we need to measure the mass of a typical galaxy and intergalactic distance. To this end, we climb up the cosmic distance ladder from the mean distance between the Earth and the Sun, 1.5×10^{13} cm, to the size of the large-scale structure of the universe, 10^{26} cm, which is an interesting and instructive subject on observational cosmology in its own right. Nevertheless, using calculated abundance of BBN, the matter density of baryons except dark matter is determined with great accuracy from the observed primordial abundance of D.

In this book, we start to make introductory explanations of General Relativity and physical processes which are necessary to construct a model of the Big Bang universe. Moreover, to explain the consistency between the calculated and observed abundances of light elements, we must explore many physical possibilities. In particular, we emphasize the importance of the idea beyond the standard BBN, because fundamental questions of both dark matter and dark energy are not answered at all.

This book is meant to become a survey of the state of physical cosmology, including observational and theoretical parts that establish the subject as a mature physical science and the more notable attempts to improve and extend the picture. The goal of this book is to present theoretical bases and applications of BBN. Readers can appreciate the current status of BBN and acquire techniques to calculate nucleosynthesis. As a result, they obtain a new idea for BBN in the framework of modern cosmology. The contents include a basic curriculum of physics, that is, mechanics, thermodynamics, quantum mechanics, and statistical physics. If a reader is unfamiliar with General Relativity, we recommend to skip complicated expressions, because he or she can comprehend the fundamental concept of BBN without the intricate formulae. Thus, most of this book is accessible to people who have studied physics to an advanced level at university; we hope that not only graduated students but also undergraduates would become familiar with BBN and receive inspiration from the early universe.

We are grateful to Katsuhiko Sato for providing his original code of BBN calculations and many useful discussions about nucleosynthesis inclusively. Thanks are due to Fraser Watson for reading through a draft of the manuscript. We also thank many graduated students who have enjoyed investigating cosmology together.

Fukuoka, Japan — Masa-aki Hashimoto
Fukuoka, Japan — Riou Nakamura
Fukuoka, Japan — E. P. Berni Ann Thushari
Kumamoto, Japan — Kenzo Arai
March 2018

Contents

Acronyms

BBN	Big Bang nucleosynthesis
BD	Brans-Dicke theory
BDΛ	Brans-Dicke model with a variable cosmological term
CL	Confidence level
CMB	Cosmic microwave background
DLA	Damped Lyman α system
COBE	Cosmic Background Explorer
SBBN	Standard Big Bang nucleosynthesis
SN Ia	Type Ia Supernova
WMAP	Wilkinson Microwave Anisotropy Probe

Chapter 1
Introduction

Abstract We make a brief historical review of the Big Bang model from A. Einstein to G. Gamow. Then we describe achievements of Big Bang nucleosynthesis (BBN) and introduce many studies on the accelerated expansion of the universe. Non-standard approaches are also exemplified with neutrino degeneracy or a decaying cosmological term. We also introduce brane-world cosmology in a five-dimensional space-time and a modified Brans-Dicke (BD) model with a variable cosmological term. Finally, we describe the structure of the present text.

Keywords Big bang · Cosmological principle · Early universe · Hubble's law · Nucleosynthesis

1.1 Historical Review

There exist a number of galaxies with various directions and distances in our real universe. Although the universe is inhomogeneous at smaller scales, it is close to homogeneous and isotropic at a sufficiently large scale, which we call the cosmological principle. A large portion of modern cosmology is based on this principle: all positions in the universe are essentially equivalent. It means that neither the Sun nor our Galaxy occupies any favored position in space.

The theory of general relativity is a geometric theory of gravitation presented in 1916 by A. Einstein [1]. He generalized Newton's law of universal gravitation to incorporate geometric property of space-time as written in a tensor form. Applying general relativity to the structure of the universe as a whole, in 1917 Einstein introduced a cosmological constant to hold back gravity and achieve a static model of the universe with homogeneous matter density [2]. In 1922, A. Friedmann was the first to show that a homogeneous universe can expand or contract [3]. In 1927, G. Lemaître rediscovered Friedmann's result and also proposed that an expanding universe accounts for the radial velocities of extragalactic nebulae [4]. In 1931 he recognized that the expansion of the universe can be traced back to a dense state in an early stage which he termed the "primeval atom" [5]. R. Tolman in 1931 remarked that radiation in a homogeneous universe relaxed to thermal

M. Hashimoto et al., *Big-Bang Nucleosynthesis*, SpringerBriefs in Physics,
https://doi.org/10.1007/978-981-13-2935-7_1

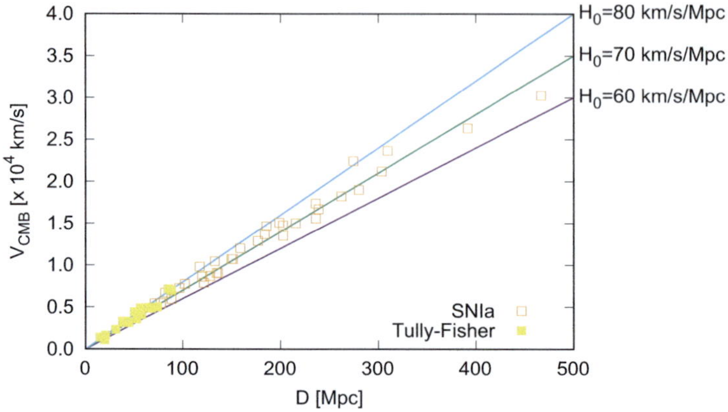

Fig. 1.1 Hubble diagram for type Ia supernovae and the Tully-Fisher relation. (Data are taken from Ref. [10])

equilibrium. He also showed that the expansion of the universe preserves the blackbody spectrum with decreasing temperature [6]. H. Robertson [7] in 1935, and independently A. Walker [8] in 1937, rigorously proved that the metric obtained by Friedmann and Lemaître is the only one on a space-time which satisfies the cosmological principle. Thereafter the Friedmann-Lemaître-Robertson-Walker, or simply Robertson-Walker, metric has been used as a fundamental framework in modern cosmology.

In 1929, E. Hubble discovered the fact that the redshift in light coming from a distant galaxy is linearly proportional to its distance [9], which is called Hubble's law: $v = H_0 D$, where v is the recession velocity, determined from the redshift, of a galaxy at a distance D and H_0 is the Hubble constant. This is strong observational evidence for the expansion of the universe. Figure 1.1 shows the Hubble diagram for type Ia supernovae (hereafter SNe Ia) and the Tully-Fisher relation, where the data are taken from the Hubble Space Telescope Key Project [10].

Hydrogen is the most abundant chemical element in the universe. The second most abundant is helium, which amounts to about 25% by mass fraction, and the remaining 2% is comprised of heavy elements. G. Gamow in the 1940s proposed that all these elements were built up by sequences of nuclear reactions in a very hot and dense phase during an early stage of the expanding universe. With his graduate student, R. Alpher, the build-up process was presented in 1948 by Alpher, Bethe, and Gamow [11]. Bethe was only added as a play on words to read the first three letters of the Greek alphabet which indicate the beginning of the world. This is referred to as the $\alpha\beta\gamma$ theory, or the Big Bang model of cosmology. The term "Big Bang" was coined with a touch of irony in 1949 by F. Hoyle, one of the proposers of steady-state cosmology, contrary to the evolving universe. It is noted that the produced amount of helium is sensitive to the neutron-to-proton ratio at the onset of nucleosynthesis. It was assumed in the $\alpha\beta\gamma$ paper that initial pure neutrons convert to protons only via β-decay of neutrons. More detailed calculations were made in

1950 by C. Hayashi [12], involving spontaneous and induced β-processes. In the $\alpha\beta\gamma$ paper, they predicted the present temperature of the relict radiation to be about 10 K, which was revised to $\simeq$5 K in 1950 by Alpher and Herman [13].

Gamow's idea of an evolving universe was completely forgotten for a while. In 1965 A. Penzias and R. Wilson reported measurement of microwave background radiation with temperature 3.5 K [14], which was accompanied with theoretical interpretation by R. Dicke et al. [15]. This radiation is the relic from the hot early universe proposed by Gamow. A modern calculation of Big Bang nucleosynthesis (hereafter BBN) was made in 1966 by P. Peebles [16, 17]. Then extensive calculations were performed with more nuclides and reactions in 1967 by R. Wagoner, W. Fowler, and F. Hoyle [18].

Nuclear reaction rates are essentially functions of density and temperature. Average baryon density of the universe at the present epoch is determined from the density of intergalactic gas and/or the product of typical mass of a galaxy and the number density of galaxies. It is noted that the density is estimated to be about 10^{-31} g cm^{-3} with uncertainty of a factor 10. On the contrary, precise measurement of the blackbody spectrum of cosmic microwave background (CMB) radiation was performed in 1989 by the COBE satellite [19]. The temperature was derived to be 2.735 ± 0.06 K. Large-scale anisotropy of CMB was also detected in temperature variations across the sky. Furthermore, small-scale fluctuations were measured in 2002 by the satellite WMAP [20], whose anisotropy spectrum was analyzed to yield cosmological parameters with great accuracy. Hence there remains baryon density to be determined from other approaches such as BBN.

1.2 Motivation

The standard model of the Big Bang based on the theory of general relativity has achieved great success in explaining the origin of light elements D, ^{4}He, and ^{7}Li. BBN has been investigated extensively along with progress in nuclear and particle physics [21–23]. However, for the last two decades, we have heard that the nature of many astronomical indications for cosmic acceleration inferred from observations of SNe Ia is one of the important challenges in particle physics. To understand the mechanism of the accelerated expansion, theoretical possibilities are proposed such as cosmography [24], brane-world cosmology [25], inhomogeneous cosmology [26, 27], and dark energy with negative pressure. It is noted that the present acceleration is due to the sort of dark energy which is represented by Einstein's cosmological term.

Observations of SNe Ia [28–30], small-scale fluctuations in CMB temperature [31, 32], and baryon acoustic oscillation [33] strongly suggest that dark energy accounts for nearly 70% of the total energy density of the universe. However the cosmological Λ term in the early universe should be some 120 orders of magnitude larger compared to the present value of about 10^{-56} cm^{-2} inferred from observations. This is called the "cosmological constant problem." It is proposed that the Λ term is not really constant but variable as time goes by. To approach this kind of problem in cosmology, we need new modified theories beyond the standard model. Various functional forms have been examined for the behavior of the Λ term. The mechanism of the dynamic reduction of the term is formulated as a time-dependent function [34] and/or represented in terms of a scalar field [35, 36]. Moreover, generalized scalar-tensor theories have been investigated [37, 38]. The evolution of the scale factor is studied with several possibilities of the functional forms [39–41].

1.3 Nonstandard Approaches to BBN

Although the standard model of BBN (SBBN) has been a great triumph, many alternatives are proposed. For example, BBN has been examined including degenerate neutrinos [18, 42] or strong density fluctuations [43–46].

There exist more baryons in our universe than antibaryons. It follows that leptons may also have large asymmetry. We examine the case of degenerate neutrinos. The background model of the universe is the same as that in the standard model, so that the dynamics is governed by the Friedmann equation. If electron-neutrinos are degenerate, the neutron-to-proton ratio alters through β-equilibrium, and consequently the produced abundance of light elements changes significantly from the standard model.

If our universe is a brane-world in a five-dimensional space-time, we take into account incoming energy from the outside and interaction energy on the brane itself. It is then expected that the expansion rate and hence the yields of BBN become different from those of the standard model.

Numerous cosmological models have been proposed with the common property of the Λ term decreasing with time. The decaying vacuum energy into CMB photons is related to the abundance of light elements [47] and the CMB intensity [48]. Our special interest is its dependence on the scale factor [39]. The interacting dark energy (or Λ term) with CMB photons has been discussed in a model with variable Λ [49, 50]. The decaying Λ term also modifies the thermal history of the early universe [51–53] and critical processes such as formation of molecules and first stars.

Along with the theory developed by Brans and Dicke [54], scalar-tensor theories are proposed by introducing the Λ term which is a function of a scalar field. BBN has been investigated [55, 56] in the Brans-Dicke (BD) model with variable Λ for the coupling constant less than or equal to 500. But the current observations [57, 58] have suggested that the constant exceeds the limit of 40,000. The parameters inherent in this model are constrained from comparison between the calculated and observed abundances of light elements [59].

1.4 Structure of This Text

In Chap. 2 the standard model of cosmology is explained based on the theory of general relativity. Since BBN depends crucially on thermonuclear reaction rates, the relevant formulae are presented for resonant and nonresonant reactions. We briefly review the observed abundances of light elements D, ^{4}He, and ^{7}Li. The calculated abundance of SBBN is compared with the primordial abundance inferred from the observed values. We discuss contribution from uncertainties in nuclear reaction rates and neutron lifetime. Then we turn to examine the magnitude-redshift relation for SNe Ia as another success of the standard model.

In Chap. 3 alternatives are introduced beyond the standard model. First, we investigate BBN with lepton asymmetry which does not change the gravitational theory but alters the physical properties of leptons. Reaction rates between neutrons and protons are briefly discussed. Second, we explore brane-world cosmology in a five-dimensional space-time which presents an extension of general relativity in connection with elementary particle physics. The Friedmann-like equation is derived to describe the expansion of the universe. Lastly, we examine an example of thermal history in the early universe with a decaying cosmological term which is treated as a part of the energy-momentum tensor in Einstein's field equation.

In Chap. 4 we investigate the modified BD theory as an alternative to general relativity. The dynamics and characteristics are described in detail for the BD model with a variable Λ term. Then we summarize observational constraints on the intrinsic parameters contained in the theory. Reasonable ranges of the inherent parameters are derived from the calculated abundance of BBN and the magnitude-redshift relation of SNe Ia.

Chapters 5, 6, 7, and 8 are Appendixes A–D, where we supplement basic formulae of tensors, thermal history during annihilation of electron-positron pairs in the early universe, a numerical method of nucleosynthesis, and physical and astronomical constants in cgs units.

References

1. Einstein, A.: Die Grundlage der Allgemeinen Relativitätstheorie. Ann. der Phys. **49**, 769–822 (1916). https://doi.org/10.1002/andp.19163540702
2. Einstein, A.: Kosmologische Betrachtung zur Allgemeinen Relativitätstheorie. Sitz. Preuss. Akad. Wiss. 142–152 (1917)
3. Friedmann, A.: Über die Krümmung des Raumes. Z. Phys. **10**, 377–386 (1922). https://doi.org/10.1007/BF01332580
4. Lemaître, G.: Un Univers Homogene de Masse Constante et de Rayon Croissant Rendant Compte de la Vitesse Radiale des Nebuleuses Extra-galactiques. Ann. Soc. Sci. Brux. A **47**, 49 (1927); English trans. A homogeneous universe of constant mass and increasing radius accounting for the radial velocity of extra-galactic nebulae. Mon. Not. R. Astron. Soc. **91**, 483–490 (1931). https://doi.org/10.1093/mnras/91.5.483

5. Lemaître, G.: The beginning of the world from the point of view of quantum theory. Nature **127**, 706 (1931). https://doi.org/10.1038/127706b0
6. Tolman, R.C.: On the problem of the entropy of the universe as a whole. Phys. Rev. **37**, 1639–1660 (1931). https://doi.org/10.1103/PhysRev.37.1639
7. Robertson, H.P.: Kinematics and world-structure. Astrophys. J. **82**, 284–301 (1935). https://doi.org/10.1086/143681
8. Walker, A.G.: On Milne's theory of world-structure. Proc. Lond. Math. Soc. Ser. 2 **42**, 90–127 (1937). https://doi.org/10.1112/plms/s2-42.1.90
9. Hubble, E.: A relation between distance and radial velocity among extra-galactic nebulae. Proc. Nat. Acad. Sci. **15**, 168–173 (1929). https://doi.org/10.1073/pnas.15.3.168
10. Freedman, W.L., et al.: Final results from the hubble space telescope key project to measure the hubble constant. Astrophys. J. **553**, 47–72 (2001). https://doi.org/10.1086/320638
11. Alpher, R.A., Bethe, H., Gamow, G.: The origin of chemical elements. Phys. Rev. **73**, 803–804 (1948). https://doi.org/10.1103/PhysRev.73.803
12. Hayashi, C.: Proton-neutron concentration ratio in the expanding universe at the stages preceding the formation of the elements. Prog. Theory Phys. **5**, 224–235 (1950). https://doi.org/10.1143/ptp/5.2.224
13. Alpher, R.A., Herman, R.C.: Theory of the origin and relative abundance distribution of the elements. Rev. Mod. Phys. **22**, 153–212 (1950). https://doi.org/10.1103/RevModPhys.22.153
14. Penzias, A.A., Wilson, R.W.: A measurement of excess antenna temperature at 4080 Mc/s. Astrophys. J. **142**, 419–421 (1965). https://doi.org/10.1086/148307
15. Dicke, R.H., Peebles, P.J.E., Roll, P.G., Wilkinson, D.T.: Cosmic black-body radiation. Astrophys. J. **142**, 414–419 (1865). https://doi.org/10.1086/148306
16. Peebles, P.J.E.: Primeval helium abundance and the primeval fireball. Phys. Rev. Lett. **16**, 410–413 (1966). https://doi.org/10.1103/PhysRevLett.16.410
17. Peebles, P.J.E.: Primordial helium abundance and the primordial fireball II. Astrophys. J. **146**, 542–552 (1966). https://doi.org/10.1086/148918
18. Wagoner, R.V., Fowler, W.A., Hoyle, F.: On the synthesis of elements at very high temperatures. Astrophys. J. **148**, 3–49 (1967). https://doi.org/10.1086/149126
19. Mather, J.C., et al.: A preliminary measurement of the cosmic microwave background spectrum by the cosmic background explorer (COBE) satellite. Astrophys. J. **354**, L37–L40 (1990). https://doi.org/10.1086/185717
20. Spergel, D.N., et al.: First-year Wilkinson microwave anisotropy probe (WMAP) observations: determination of cosmological parameters. Astrophys. J. Suppl. **148**, 175–194 (2003). https://doi.org/10.1086/377226
21. Schramm, D.N., Wagoner, R.V.: Element production in the early universe. Ann. Rev. Nucl. Sci. **27**, 37–74 (1977)
22. Boesgaard, A.M., Steigman, G.: Big bang nucleosynthesis: theories and observations. Ann. Rev. Astron. Astrophys. **23**, 319–378 (1985)
23. Mathews, G.J., Kusakabe, M., Kajino, T.: Introduction to big bang nucleosynthesis and modern cosmology. Int. J. Mod. Phys. E **26**, 1741001 (2017). https://doi.org/10.1142/S0218301317410014
24. Rapetti, D., Allen, S., Amin, M., Blandford, R.: A kinematical approach to dark energy studies. Mon. Not. R. Astron. Soc. **375**, 1510–1520 (2007). https://doi.org/10.1111/j.1365-2966.2006.11419.x
25. Umezu, K.I., Ichiki, K., Kajino, T., Mathews, G.J., Nakamura, R., Yahiro, M.: Observational constraints on accelerating brane cosmology with exchange between the bulk and brane. Phys. Rev. D **73**, 063527 (2006). https://doi.org/10.1103/PhysRevD.73.063527
26. Alnes, H., Amarzguioui, M., Groen, O.: An inhomogeneous alternative to dark energy? Phys. Rev. D **73**, 083519 (2006). https://doi.org/10.1103/PhysRevD.73.083519
27. Romano, A.E.: Lemaitre-Tolman-Bondi universes as alternatives to dark energy: does positive averaged acceleration imply positive cosmic acceleration? Phys. Rev. D **75**, 043509 (2007). https://doi.org/10.1103/PhysRevD.75.043509

28. Riess, A.G., et al.: Observational evidence from supernovae for an accelerating universe and a cosmological constant. Astron. J. **116**, 1009–1038 (1998). https://doi.org/10.1086/300499
29. Perlmutter, S., et al.: Measurements of Ω and Λ from 42 high-redshift supernovae. Astrophys. J. **517**, 565–586 (1999). https://doi.org/10.1086/307221
30. Amanullah, R., et al.: Spectra and light curves of six type Ia supernovae at $0.511 < z < 1.12$ and the Union2 compilation. Astrophys. J. **716**, 712–738 (2010). https://doi.org/10.1088/0004-637X/716/1/712
31. Brown, M.L., et al.: Improved measurements of the temperature and polarization of the CMB from QUaD. Astrophys. J. **705**, 978–999 (2009). https://doi.org/10.1088/0004-637X/705/1/978
32. Komatsu, E., et al.: Seven-year Wilkinson microwave anisotropy probe (WMAP) observations: cosmological interpretation. Astrophys. J. Suppl. **192**, 18 (2011). https://doi.org/10.1088/0067-0049/192/2/18
33. Percival, W.J., et al.: Baryon acoustic oscillations in the sloan digital sky survey data release 7 galaxy sample. Mon. Not. R. Astron. Soc. **401**, 2148–2168 (2010). https://doi.org/10.1111/j.1365-2966.2009.15812.x
34. Silveira, V., Waga, I.: Cosmological properties of a class of Λ decaying cosmologies. Phys. Rev. D **56**, 4625–4632 (1997). https://doi.org/10.1103/PhysRevD.56.4625
35. Weinberg, S.: The cosmological constant problem. Rev. Mod. Phys. **61**, 1–23 (1989). https://doi.org/10.1103/RevModPhys.61.1
36. Huterer, D., Turner, M.S.: Prospects for probing the dark energy via supernova distance measurements. Phys. Rev. D **60**, 081301 (1999). https://doi.org/10.1103/PhysRevD.60.081301
37. Wagoner, R.V.: Scalar-tensor theory and gravitational waves. Phys. Rev. D **1**, 3209–3216 (1970). https://doi.org/10.1103/PhysRevD.1.3209
38. Endo, M., Fukui, T.: The cosmological term and a modified Brans-Dicke cosmology. Gen. Rel. Grav. **8**, 833–839 (1977). https://doi.org/10.1007/BF00759587
39. Kimura, K., Hashimoto, M., Sakoda, M., Arai, K.: Effects on the temperatures of a variable cosmological term after recombination. Astrophys. J. **561**, L19–L22 (2001). https://doi.org/10.1086/324569
40. Hashimoto, M., Kamikawa, T., Arai, K.: Effects of a decaying cosmological term on the formation of molecules and first objects. Astrophys. J. **598**, 13–19 (2003). https://doi.org/10.1086/378882
41. Solà, J., Štefančić, H.: Dynamical dark energy or variable cosmological parameters? Mod. Phys. Lett. A **21**, 479–494 (2006). https://doi.org/10.1142/S0217732306019554
42. Yahil, A., Beaudet, G.: Big-bang nucleosynthesis with nonzero lepton numbers. Astrophys. J. **206**, 26–29 (1976). https://doi.org/10.1086/154352
43. Applegate, J.H., Hogan, C.J., Scherrer, R.J.: Cosmological baryon diffusion and nucleosynthesis. Phys. Rev. D **35**, 1151–1160 (1987). https://doi.org/10.1103/PhysRevD.35.1151
44. Terasawa, N., Sato, K.: Neutron diffusion and nucleosynthesis in the universe with isothermal fluctuations produced by Quark-Hadron phase transition. Phys. Rev. D **39**, 2893–2900 (1989). https://doi.org/10.1103/PhysRevD.39.2893
45. Nakamura, R., Hashimoto, M., Fujimoto, S., Sato, K.: Constraint on heavy element production in inhomogeneous big-bang nucleosynthesis from the light element observations. J. Astrophys. **2013**, 587294 (2013). https://doi.org/10.1155/2013/587294
46. Nakamura, R., Hashimoto, M., Ichimasa, R., Arai, K.: Big bang nucleosynthesis: constraints on nuclear reaction rates, neutrino degeneracy, inhomogeneous and Brans-Dicke models. Int. J. Mod. Phys. E **26**, 1741003 (2017). https://doi.org/10.1142/S0218301317410038
47. Freese, K., Adams, F.C., Frieman, J.A., Mottola, E.: Cosmology with decaying vacuum energy. Nucl. Phys. B **287**, 797–814 (1987). https://doi.org/10.1016/0550-3213(87)90129-5
48. Overduin, J.M., Wesson, P.S., Bowyer, S.: Constraints on vacuum decay from the microwave background. Astrophys. J. **404**, 1–7 (1993). https://doi.org/10.1086/172253
49. Puy, D.: Thermal balance in decaying Λ cosmologies. Astron. Astrophys. **422**, 1–9 (2004). https://doi.org/10.1051/0004-6361:20040256

50. Nakamura, R., Hashimoto, M., Ichiki, K.: Cosmic microwave background constraints on a decaying cosmological term related to the thermal evolution. Phys. Rev. D **77**, 123511 (2008). https://doi.org/10.1103/PhysRevD.77.123511
51. Lima, J.A.S.: Thermodynamics of decaying vacuum cosmologies. Phys. Rev. D **54**, 2571–2577 (1996). https://doi.org/10.1103/PhysRevD.54.2571
52. Jetzer, P., Puy, D., Signore, M., Tortora, C.: Limits on decaying dark energy density models from the CMB temperature-redshift relation. Gen. Relativ. Gravit. **43**, 1083–1093 (2011). https://doi.org/10.1007/s10714-010-1091-4
53. Jetzer, P., Tortora, C.: Constraints from the CMB temperature and other common observational data-sets on variable dark energy density models. Phys. Rev. D **84**, 043517 (2011). https://doi.org/10.1103/PhysRevD.84.043517
54. Brans, C., Dicke, R.H.: Mach's principle and a relativistic theory of gravitation. Phys. Rev. **124**, 925–935 (1961). https://doi.org/10.1103/PhysRev.124.925
55. Arai, K., Hashimoto, M., Fukui, T.: Primordial nucleosynthesis in the Brans-Dicke theory with a variable cosmological term. Astron. Astrophys. **179**, 17–22 (1987)
56. Etoh, T., Hashimoto, M., Arai, K., Fujimoto, S.: Age of the universe constrained from the primordial nucleosynthesis in the Brans-Dicke theory with a varying cosmological term. Astron. Astrophys. **325**, 893–897 (1997)
57. Bertotti, B., Iess, L., Tortora, P.: A test of general relativity using radio links with the cassini spacecraft. Nature **425**, 374–376 (2003). https://doi.org/10.1038/nature01997
58. Berti, E., Buonanno, A., Will, C.M.: Estimating spinning binary parameters and testing alternative theories of gravity with LISA. Phys. Rev. D **71**, 084025 (2005). https://doi.org/10.1103/PhysRevD.71.084025
59. Nakamura, R., Hashimoto, M., Gamow, S., Arai, K.: Big-bang nucleosynthesis in a Brans-Dicke cosmology with a varying Λ term related to WMAP. Astron. Astrophys. **448**, 23–27 (2006). https://doi.org/10.1051/0004-6361:20042618

Chapter 2
The Standard Model of Cosmology

Abstract We mention the fundamentals in the theory of general relativity, i.e., the principles of equivalence and general covariance. Based on the facts that our universe is homogeneous and isotropic at large scales, we derive the Robertson-Walker metric and subsequently the Friedmann equation which governs the expansion of the universe. The equation is expressed in a convenient form in terms of density parameters. Next, we describe thermonuclear reaction rates utilized in astrophysics, which involve resonant and nonresonant reactions, photodisintegration, electron capture, and β-decay. We review the significance of standard Big Bang nucleosynthesis (SBBN) and summarize the current situation in the observed primordial abundance of light elements, ^{4}He, D, and ^{7}Li. Comparing the calculated abundance of the elements with observed values, we determine a reasonable range for the baryon-to-photon ratio. Moreover, we examine the dependence of the produced amount of ^{4}He on the measured lifetimes of neutrons. Finally, we consider the magnitude-redshift relation of type Ia supernovae (SNe Ia) as an independent probe to cosmological models.

Keywords Dark energy · Dark matter · Friedmann equation · General relativity · Magnitude-redshift relation · Primordial abundance · Thermonuclear reaction

2.1 Theory of General Relativity

In this chapter, we adopt the theory of general relativity which is now the standard theory of cosmology. General relativity is a geometric theory of gravitation. When there exists an object of large mass, its surrounding space is not flat but concave. Consider a test particle approaching the object. Its trajectory is curved toward the object. In a Newtonian scheme, the particle is attracted by the gravitational force of the object. According to Einstein, the particle takes a minimum path, i.e., a geodesic in a curved space. More typically let us consider a ray of light which is a massless particle. The light proceeds along a straight line in a Newtonian picture, but it takes

M. Hashimoto et al., *Big-Bang Nucleosynthesis*, SpringerBriefs in Physics,
https://doi.org/10.1007/978-981-13-2935-7_2

also a minimum path, which is curved, in general relativity. In an extreme case, light is totally trapped and consequently disappears from sight.

General relativity is founded on the following two principles. One is the principle of equivalence. It tells that gravitational and inertial forces are indistinguishable in a small region. It is exemplified by a famous thought experiment of an elevator. Note that vertical trajectories of two bodies, which fall freely on the Earth, are parallel only in a small region. They gradually become closer in a large scale. The other is the principle of general covariance. It demands that a physical law is covariant under general transformation of coordinates, so that it should be represented in a tensor form.

The distance ds between neighboring points x^μ and $x^\mu + dx^\mu$ is given by

$$ds^2 = g_{\mu\nu} dx^\mu dx^\nu, \tag{2.1}$$

where $g_{\mu\nu}$ is the metric tensor which describes geometrical properties of four-dimensional space-time and the Greek letter runs from 0 to 3. The repeated upper and lower suffixes are summed over the range.

Being analogous to Poisson's equation for Newtonian gravitational potential, Einstein's field equation is written as [1]

$$R_{\mu\nu} - \frac{1}{2} g_{\mu\nu} R + \Lambda g_{\mu\nu} = 8\pi G T_{\mu\nu}. \tag{2.2}$$

Here $R_{\mu\nu}$ is the Ricci tensor, R is the curvature scalar, Λ is the cosmological constant, $T_{\mu\nu}$ is the energy momentum tensor that measures the relevant properties of the constituent in the universe, and G is the gravitational constant. Note that we adopt the units where the light velocity $c = 1$ throughout. Expressions of relevant tensors are summarized in Appendix A.

2.2 The Robertson-Walker Metric

The universe is observed to be close to homogeneous and isotropic at large scales. This means that we stand on neither preferred center nor edge of the universe and what we see looks the same in any direction. This is called the cosmological principle. The three-dimensional spaces at any constant time are locally isotropic. Then, the spatial part of the line element is written in terms of the spherical coordinates r, θ, and φ as

$$d\sigma^2 = dr^2 + r^2 d\theta^2 + r^2 \sin^2\theta \, d\varphi^2.$$

The complete line element in our space-time is given in a general form by

$$ds^2 = -dt^2 + \exp[f(t, r)] \, d\sigma^2, \tag{2.3}$$

where the function f does not depend on θ and φ because of the isotropy. We take the signature $(-1, 1, 1, 1)$. Cross terms such as $dt\,dr$ disappear due to the homogeneity.

The nonvanishing Christoffel symbols (5.2) are written as follows:

$$\begin{aligned}
&\Gamma^0_{11} = e^{2f}\dot{f}, \quad \Gamma^0_{22} = e^{2f}r^2\dot{f}, \quad \Gamma^0_{33} = {\Gamma^0}_{22}\sin^2\theta, \\
&\Gamma^1_{01} = \dot{f}, \quad \Gamma^1_{11} = f', \quad \Gamma^1_{22} = -(f'r^2 + r), \\
&\Gamma^1_{33} = \Gamma^1_{22}\sin^2\theta, \quad \Gamma^2_{02} = \Gamma^3_{03} = \dot{f},\ \Gamma^2_{12} = \Gamma^3_{13} = f' + \frac{1}{r}, \\
&\Gamma^2_{33} = -\sin\theta\cos\theta,\ \Gamma^3_{23} = \cot\theta,
\end{aligned}$$

where the dot and prime denote the differentiation with respect to t and r, respectively. The Ricci tensors (5.8) are

$$\begin{aligned}
R_{00} &= -3\left(\ddot{f} + \dot{f}^2\right), \\
R_{01} &= -2\dot{f}', \\
R_{11} &= e^{2f}\left(\ddot{f} + 3\dot{f}^2\right) - 2\left(f'' + \frac{1}{r}f'\right), \\
R_{22} &= e^{2f}r^2\left(\ddot{f} + 3\dot{f}^2\right) - \left(f''r^2 + f'^2r^2 + 3f'r\right), \\
R_{33} &= R_{22}\sin^2\theta.
\end{aligned}$$

Thus the curvature scalar (5.9) is

$$R = 6\left(\ddot{f} + 2\dot{f}^2\right) - 2e^{-2f}\left(2f'' + f'^2 + \frac{4}{r}f'\right). \tag{2.4}$$

We, therefore, obtain the Einstein tensors (5.10)

$$G^0{}_0 = -3\dot{f}^2 + e^{-2f}\left(2f'' + f'^2 + \frac{4}{r}f'\right), \tag{2.5}$$

$$G^1{}_1 = -\left(2\ddot{f} + 3\dot{f}^2\right) + e^{-2f}\left(f'^2 + \frac{2}{r}f'\right), \tag{2.6}$$

$$G^2{}_2 = G^3{}_3 = -\left(2\ddot{f} + 3\dot{f}^2\right) + e^{-2f}\left(f'' + \frac{1}{r}f'\right), \tag{2.7}$$

$$G^0{}_1 = 2\dot{f}'. \tag{2.8}$$

The energy-momentum tensor can be written as the form of a perfect fluid:

$$T_{\mu\nu} = (\rho + p)U_\mu U_\nu + p g_{\mu\nu}, \tag{2.9}$$

where ρ is the energy density, p is the pressure and U_μ is the four velocity. Since $U_\mu = (-1, 0, 0, 0)$ for a particle at rest in this frame, we get

$$T^0{}_0 = -\rho, \quad T^1{}_1 = T^2{}_2 = T^3{}_3 = p.$$

From (2.8) the condition $G^0{}_1 = 0$ leads to

$$\dot{f}' = 0.$$

The variables t and r can be separated to be

$$f = F(r) + \ln a(t). \tag{2.10}$$

Furthermore, from (2.6) and (2.7), the condition $G^1{}_1 = G^2{}_2$ yields

$$f'' + \frac{1}{r} f' = f'^2 + \frac{2}{r} f'.$$

With use of (2.10), we have

$$F'' - F'^2 - \frac{1}{r} F' = 0.$$

If we set $F' = r\Phi(r)$, we get

$$r\Phi' - r^2\Phi^2 = 0.$$

It follows that

$$-\frac{1}{\Phi} = \frac{1}{2} r^2 + \frac{2}{k},$$

where k is a constant. Then we obtain

$$F' = -\frac{kr}{2 + kr^2/2},$$

which is integrated to be

$$F = -\ln\left(1 + \frac{1}{4} kr^2\right).$$

We get from (2.10)

$$\mathrm{e}^f = \frac{a}{1 + kr^2/4}. \tag{2.11}$$

Therefore we obtain the line element (2.3) as

$$ds^2 = -dt^2 + \frac{a^2(t)}{(1 + kr^2/4)^2}\left(dr^2 + r^2 d\theta^2 + r^2 \sin^2\theta d\varphi^2\right), \tag{2.12}$$

where $a(t)$ is the scale factor, which describes the expansion of the universe. The dimension of length is carried by $a(t)$, so that the radial coordinate r is dimensionless. The expression (2.12) is called the Robertson-Walker metric. Since dynamics of the universe is expressed in terms of $a(t)$, galaxies at rest in the coordinate system (r, θ, φ) continue to remain at rest even if the universe expands. Hence this is named the comoving coordinate system which follows the motion of typical galaxies.

Inserting (2.11) into the second term of (2.4) and neglecting the time variation, we get the curvature scalar

$$R = \frac{6k}{a^2}. \tag{2.13}$$

It follows that the curvature is constant in Robertson-Walker space-time. The three-dimensional space of $t =$ constant is flat for $k = 0$, closed with positive curvature for $k = +1$, or open with negative curvature for $k = -1$. Therefore, k is called the curvature constant.

When we introduce a new radial coordinate

$$\tilde{r} = \frac{r}{1 + kr^2/4},$$

then we have

$$d\tilde{r} = \frac{1 - kr^2/4}{(1 + kr^2/4)^2} dr.$$

If we rewrite r instead of $\tilde{r}$, then (2.12) becomes

$$ds^2 = -dt^2 + a^2(t)\left(\frac{dr^2}{1 - kr^2} + r^2 d\theta^2 + r^2 \sin^2\theta d\varphi^2\right). \tag{2.14}$$

This is an alternative representation of the Robertson-Walker metric.

Equation (2.14) can be expressed in another useful form

$$ds^2 = -dt^2 + a^2(t)[d\chi^2 + S^2(\chi)(d\theta^2 + \sin^2\theta d\varphi^2)]$$

with

$$S(\chi) = \begin{cases} \sin\chi & (k = +1), \\ \chi & (k = 0), \\ \sinh\chi & (k = -1). \end{cases}$$

2.3 The Friedmann Equation

The conservation law of energy-momentum, i.e., the condition that the covariant divergence of the energy-momentum tensor vanishes, is written as

$$T^{\mu}{}_{\nu;\mu} = \frac{\partial T^{\mu}{}_{\nu}}{\partial x^{\mu}} + \Gamma^{\mu}_{\mu\lambda} T^{\lambda}{}_{\nu} - \Gamma^{\lambda}_{\mu\nu} T^{\mu}{}_{\lambda} = 0,$$

where the semicolon denotes the covariant derivative (5.6). Using (2.9), the $\nu = 0$ component is the energy conservation law:

$$\frac{d}{dt}\left(\rho a^3\right) + p\frac{da^3}{dt} = 0. \tag{2.15}$$

If an equation of state is set to

$$p = (\gamma - 1)\rho, \tag{2.16}$$

where γ is a constant, then (2.15) reduces to

$$\frac{d}{dt}\left(\rho a^{3\gamma}\right) = 0,$$

which is integrated to

$$\rho a^{3\gamma} = \text{const.} \tag{2.17}$$

For baryons and dark matter we set $\gamma = 1$, i.e., $p_m = 0$, then (2.17) becomes

$$\rho_m a^3 = \rho_{m0} a_0^3 = \text{const,} \tag{2.18}$$

where the subscript zero denotes the value at the present epoch. Similarly, for radiation and mass-less particles with $\gamma = 4/3$, we have

$$\rho_r a^4 = \rho_{r0} a_0^4 = \text{const.} \tag{2.19}$$

The energy density for a mixture of matter and radiation is expressed as

$$\begin{aligned} \rho &= \rho_m + \rho_r \\ &= \rho_{m0} a^{-3} + \rho_{r0} a^{-4}. \end{aligned} \tag{2.20}$$

For convenience, we have normalized the scale factor a to unity at the present epoch ($a_0 = 1$).

With use of the metric (2.14), the (00) component of (2.2) gives

$$H^2 = \left(\frac{\dot{a}}{a}\right)^2 = \frac{8\pi G}{3}(\rho_m + \rho_r) - \frac{k}{a^2} + \frac{\Lambda}{3}. \tag{2.21}$$

This is called the Friedmann equation which governs the expansion of the universe.

Let us introduce density parameters

$$\begin{aligned} \Omega_m &= \frac{\rho_{m0}}{\rho_{\rm cr}}, & \Omega_r &= \frac{\rho_{r0}}{\rho_{\rm cr}}, \\ \Omega_k &= -\frac{k}{H_0{}^2}, & \Omega_\Lambda &= \frac{\Lambda}{3H_0{}^2}, \end{aligned} \tag{2.22}$$

where H_0 is the Hubble constant and $\rho_{\rm cr} = 3H_0{}^2/(8\pi G)$ is the critical density for which the universe is spatially flat in a model with $\Lambda = 0$. Equation (2.21) can be written, with use of (2.22), as

$$H^2 = H_0{}^2 \left[\Omega_r a^{-4} + \Omega_m a^{-3} + \Omega_k a^{-2} + \Omega_\Lambda\right]. \tag{2.23}$$

The density parameters satisfy the following condition evaluated at the present epoch:

$$\Omega_r + \Omega_m + \Omega_k + \Omega_\Lambda = 1.$$

It is noted that radiation temperature varies as $T \sim a^{-1}$ because $\rho_r \sim a^{-4}$ from (2.19) and $\rho_r \sim T^4$. In an early universe of high temperatures $T > 10^{11}$ K, neutrinos and photons are in thermal equilibrium. Then we have $T = T_\nu \sim a^{-1}$, where T_ν is the neutrino temperature. Neutrinos decouple from photons at $T \simeq 1.7\times 10^{10}$ K. During the stage of pair annihilation of electrons and positrons at around $T \simeq 1.2\times 10^{10}$ K, energies of $e^\pm$ pairs are converted into photons. We can follow the descent in the photon temperature numerically as explained in Appendix B. After the complete pair annihilation, photon temperature is given by $T = (11/4)^{1/3} T_\nu \sim a^{-1}$.

The number density of photons is given by

$$n_\gamma = \frac{16\pi\zeta(3)}{h^3}(k_{\rm B}T)^3, \tag{2.24}$$

where $\zeta(3)$ is the zeta function.

Using (2.23), the age of the universe is given by

$$t_0 = \int_0^{t_0} dt = \int_0^1 \frac{da}{aH}. \tag{2.25}$$

Table 2.1 Cosmological parameters from WMAP 9 year [2] and *Planck* [3]

Description	Symbol	WMAP9	*Planck*
Age of universe (Gyr)	t_0	13.74 ± 0.11	13.813 ± 0.0038
Hubble constant	h	0.700 ± 0.022	0.6731 ± 0.0096
Baryon density	$\Omega_b h^2$	0.02264 ± 0.00050	0.02222 ± 0.00023
Dark matter density	$\Omega_c h^2$	0.1138 ± 0.0045	0.1197 ± 0.0022
Dark energy density	Ω_Λ	0.721 ± 0.025	0.685 ± 0.013

There are key observational quantities of the standard model. Measurements of fluctuations in CMB temperature single out the Big Bang model as the prime and promising candidate to describe our universe. The cosmological parameters are determined from the satellites WMAP [2] and *Planck* [3] with amazing accuracy. They are summarized in Table 2.1, where $h = H_0/(100\,\mathrm{kms^{-1}Mpc^{-1}})$. Note that matter is composed of baryons, or nucleons, and dark matter, i.e., $\Omega_m = \Omega_b + \Omega_c$, where the subscript "c" indicates cold dark matter (see below). The other parameter is $\Omega_r = 4.15 \times 10^{-5}\,h^{-2}$, which includes the contribution from three species of neutrinos, derived from the CMB temperature of $T_0 = 2.725 \pm 0.002\,\mathrm{K}$. We can set $\Omega_k = 0$ because of a flat space.

It can be seen from Table 2.1 that our universe is composed of 72% dark energy (cosmological constant) and 23% dark matter of energy density. Ordinary baryonic matter amounts to only 5%.

It is worthwhile to mention the characteristics of dark matter. The rotation curve of a typical spiral galaxy, including our Galaxy, exhibits that the rotation velocity is kept nearly constant to the outer parts of the galaxy. This is contrary to the well-known Kepler's law: the orbital speed u of a planet varies as $u \sim r^{-1/2}$, where r is its orbital radius. It means that the motion of the planet is governed by the gravitation of the point mass, i.e., the Sun. On the other hand, the constant rotation velocity in the galaxy indicates that a substantial mass of invisible material surrounds the visible whole galaxy. It was formerly called "missing mass" but now is termed "dark matter." Moreover, concerning the dynamics of rich clusters of galaxies, we have evidence [4] that the peculiar velocities, apart from the recession velocity due to cosmic expansion, of individual galaxies are too large to be gravitationally confined within the cluster. The estimated mass from the virial theorem is more than ten times greater than the total mass of luminous galaxies. Considerable amounts of dark matter are needed to hold clusters together.

Dark matter is a gas of particles that moves freely with null electric charge. They cannot be baryons, because high baryon density would contradict the successful theory for the origin of light elements as will be described in Sect. 2.5. If the non-baryonic particles are neutrinos with a small mass, they would be sufficiently hot and hence contribute to the energy density in the early universe. Strong restrictions are placed from BBN, again. Consequently, the dark matter must be cold and does not have nongravitational interactions with ordinary constituents of the universe. Several candidates of cold dark matter have been proposed so far: neutralino, one of weakly interacting massive particles, and axions.

2.4 Astronuclear Reactions

2.4.1 Thermonuclear Reaction Rates

We deal with nuclear fusion in a completely ionized plasma of matter density ρ and temperature T. Number densities are too low at the BBN stage for nuclei to be built up directly through many body collisions like $2\mathrm{p} + 2\mathrm{n} \longrightarrow {}^4\mathrm{He} + \gamma$. Therefore, we consider a two-body reaction between nuclei with number densities n_1 and n_2. The cross section for the reaction is a function of the relative velocity v. The reaction rate in units of volume and time is defined as follows [5]:

$$n_1 n_2 \langle \sigma v \rangle = n_1 n_2 4\pi \left(\frac{\mu}{2\pi k_\mathrm{B} T} \right)^{3/2} \int_0^\infty v^3 \sigma(v) \exp\left[-\frac{\mu v^2}{2k_\mathrm{B} T} \right] dv, \tag{2.26}$$

where k_B is the Boltzmann constant. The average value $\langle \sigma v \rangle$ is given with the aid of Maxwell's velocity distribution. The reduced mass μ for the particle masses M_1 and M_2 is

$$\mu = \frac{M_1 M_2}{M_1 + M_2}.$$

Let $E = \mu v^2/2$ be the kinetic energy in the frame of the center of mass. Then (2.26) can be written as

$$\langle \sigma v \rangle = \frac{4}{(2\pi\mu)^{1/2} (k_\mathrm{B} T)^{3/2}} \int_0^\infty E \sigma(E) \exp\left[-\frac{E}{k_\mathrm{B} T} \right] dE. \tag{2.27}$$

Therefore, if we have knowledge of the cross section $\sigma(E)$, we can derive the reaction rate as a function of T by performing integration in (2.27). When we solve simultaneous rate equations concerning abundance of nuclei as described in Appendix C, it is convenient to define the thermonuclear reaction rate as $N_\mathrm{A}\langle \sigma v \rangle$, where N_A is Avogadro's number. As a consequence, (2.27) leads to

$$N_\mathrm{A} \langle \sigma v \rangle = 3.7313 \times 10^{10} A^{-1/2} T_9^{-3/2} \times \int_0^\infty E \sigma(E) \exp\left[-\frac{11.6045}{T_9} \right] dE \quad \mathrm{cm^3 mole^{-1} s^{-1}}. \tag{2.28}$$

Here A is the reduced mass number

$$A = \frac{A_1 A_2}{A_1 + A_2} = \frac{\mu}{M_u},$$

where M_u is the atomic mass and $T_9 = T/10^9$ K. We have

$$k_B T = T_9/11.6045 \quad \text{MeV},$$

(1 MeV $= 1.6022 \times 10^{-6}$ erg).

Usually, thermonuclear reactions occur in astrophysical environment of $T_9 < 10$. For example, BBN proceeds in the range $0.01 < T_9 < 10$ and helium burning ignites in a stellar core at around $T_9 \simeq 0.1$. On the other hand, the Coulomb energy between two nuclei at contact is

$$V_C = \frac{Z_1 Z_2 e^2}{R} = \frac{1.44 Z_1 Z_2}{R} \quad \text{MeV},$$

where $R = 1.4A^{1/3}$ fm is the nuclear radius (1 fm $= 10^{-13}$ cm). Since V_C is beyond several MeV, the nuclear reaction does not occur from a viewpoint of classical mechanics. However, the reaction proceeds gradually due to the tunnel effect of quantum mechanics. The cross section is proportional to the penetration probability. In particular, it is proportional to the geometrical cross section, i.e., square of the de Broglie wavelength for low-energy nuclear reactions. As a consequence, it is convenient to write $\sigma(E)$ in the following factorized form:

$$\sigma(E) = \frac{S(E)}{E} \exp\left[-\frac{2\pi Z_1 Z_2 e^2}{\hbar v}\right],$$

where $\hbar = h/2\pi$ is the reduced Planck constant. When the interaction energy of two nuclei is not nearly equal to an energy at which the two nuclei resonate in a quasistationary state, the factor $S(E)$ is constant, or at least a slowly varying function of E. This $S(E)$ is called the astrophysical S-factor.

Therefore, we obtain from (2.27)

$$\langle \sigma v \rangle = \frac{4}{(2\pi\mu)^{1/2}(k_B T)^{3/2}} \int_0^\infty S(E) \exp\left(-\frac{E}{k_B T} - bE^{-1/2}\right) dE, \tag{2.29}$$

where

$$b = \frac{(2\mu)^{1/2}\pi Z_1 Z_2 e^2}{\hbar} = 31.29 Z_1 Z_2 A^{1/2} \quad \text{keV}^{1/2}.$$

It is noted that the thermonuclear reaction rate depends crucially on the environment whether there occurs resonance on to an energy level of the compound nucleus.

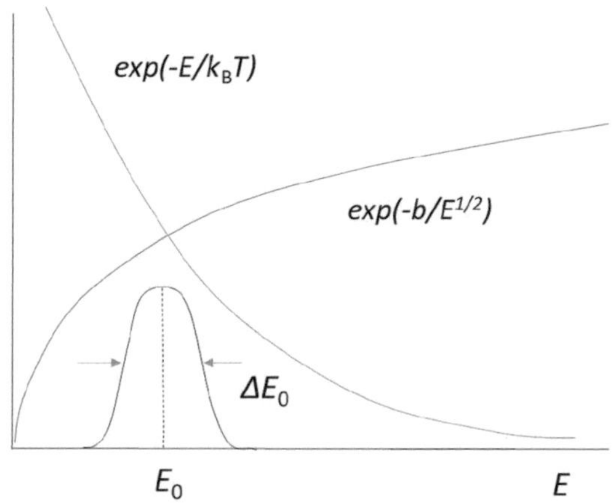

Fig. 2.1 Schematic representation of a Gamow peak

2.4.2 *Nonresonant Reactions*

We consider the case when the reaction occurs in an energy range far from the resonance energy. If $S(E)$ is nearly constant, we can set $S(E) = S_0$. The integrand in (2.29) is divided into $g_1 = \exp(-E/k_B T)$ and $g_2 = \exp(-bE^{-1/2})$. The schematic features of g_1 and g_2 are shown in Fig. 2.1. The product has a sharp maximum, which is called the "Gamow peak" at $E = E_0$ and an e-folding width of ΔE_0.

We perform a Taylor expansion of the integrand around E_0 and retain the terms up to the second order. Then we obtain

$$\exp\left(-\frac{E}{k_B T} - bE^{-1/2}\right) = \exp\left(-\frac{3E_0}{k_B T}\right)\exp\left[-\left(\frac{E-E_0}{\Delta E_0}\right)^2\right] \tag{2.30}$$

with

$$E_0 = \left(\frac{1}{2}bk_B T\right)^{2/3} = 0.122(Z_1^2 Z_2^2 A)^{1/3} T_9^{2/3} \quad \text{MeV},$$

$$\Delta E_0 = 2\left(\frac{1}{3}k_B T E_0\right)^{1/2} = 0.237(Z_1^2 Z_2^2 A)^{1/6} T_9^{5/6} \quad \text{MeV}.$$

Note that E_0 is generally much larger than $k_B T$ in astrophysical environment. Consequently (2.29) is integrated to be

$$\langle \sigma v \rangle = 1.30 \times 10^{-17}\left(\frac{Z_1 Z_2}{A}\right)^{1/3} S_0 T_9^{-2/3} e^{-\tau} \quad \text{cm}^3\text{s}^{-1},$$

where $\tau = 4.25(AZ_1^2 Z_2^2/T_9)^{1/3}$. This is the rate of the nonresonant reaction.

Table 2.2 Astrophysical S-factor extrapolated to $E = 0$ with use of NACRE-II [6]

Reaction	$S(0)$ (eV barns)
^{2}H(p,γ)^{3}He	0.21 ± 0.04
^{2}H(d,γ)^{4}He	$5.8^{+1.0}_{-1.5} \times 10^{-3}$
^{2}H(d,n)^{3}He	$5.55 \pm 0.6 \times 10^4$
^{2}H(d,p)^{3}H	$56.2^{+4.9}_{-4.7} \times 10^3$
^{2}H(α, γ)^{6}Li	$2.2^{+0.9}_{-1.2} \times 10^{-3}$
^{3}H(d,n)^{4}He	$11 \pm 1 \times 10^6$
^{3}H(α, γ)^{7}Li	98^{+11}_{-8}
^{3}He(d, p)^{4}He	$5.9 \pm 0.5 \times 10^6$
^{3}He(^{3}He,2p)^{4}He	$5.3 \pm 0.5 \times 10^6$
^{3}He(α, γ)^{7}Be	$0.56^{+0.05}_{-0.07} \times 10^3$
^{6}Li(p,γ)^{7}Be	73^{+56}_{-11}
^{6}Li(p,α)^{3}He	$3.1 \pm 0.4 \times 10^6$
^{7}Li(p,γ)^{8}Be	$1.3^{+0.4}_{-0.2} \times 10^3$
^{7}Li(p,α)^{4}He	$52^{+11}_{-8} \times 10^3$

For an astrophysical use, we express the thermonuclear reaction rate as a function of T_9:

$$N_{\rm A}\langle\sigma v\rangle = 7.83 \times 10^9 \left(\frac{Z_1 Z_2}{A}\right)^{1/3} S_0 T_9^{-2/3} {\rm e}^{-\tau} \quad {\rm cm^3 mole^{-1} s^{-1}}. \tag{2.31}$$

Table 2.2 gives values of $S(E)$ extrapolated to $E = 0$. The rates are taken from Ref. [6]. As can be seen, the magnitude of order eV barns indicates how small the reaction rates are (1 barn $= 10^{-24}$ cm^2). It is noted that, when $T_9 < 0.01$, even hydrogen burning is difficult to operate in a stellar core.

2.4.3 Resonant Reactions

When resonances occur near the Gamow peak, the cross section becomes many orders of magnitude greater than that of nonresonant reaction. The resonant reaction occupies the main part of the reaction rate, so that we consider energy dependence of the resonant cross section.

If the reaction occurs through an energy state of the compound nucleus, the cross section is written by the Breit-Wigner single-level formula with the resonance energy E_r as

$$\sigma(E) = \frac{\pi\hbar^2}{2\mu E} \frac{\omega\Gamma_a\Gamma_b}{(E - E_r)^2 + (\Gamma/2)^2}. \tag{2.32}$$

Here Γ_a and Γ_b are the partial widths of the formation and destruction of the compound nucleus, respectively, and $\Gamma = \Gamma_a + \Gamma_b$ is the total width. The statistical weight ω is given by

$$\omega = \frac{2J+1}{(2j_1+1)(2j_2+1)},$$

where j_1, j_2 and J are the spins of the projectile, target and compound nuclei, respectively. The partial width of a state to a particle emission for orbital quantum number l is written as

$$\Gamma_l = \frac{3\hbar v}{R} P_l \theta_l^2, \tag{2.33}$$

where θ_l is the dimensionless reduced width which could be known experimentally in the range of $0.01 < \theta_l^2 < 1$. The interaction radius or nuclear radius is approximately the sum of the radii of the two particles: $R = 1.4A_1^{1/3} + 1.4A_2^{1/3}$ fm. The penetration factor is given by

$$P_l = \frac{1}{F_l^2(kR) + G_l^2(kR)},$$

where F_l and G_l are the regular and irregular Coulomb wave functions, respectively, and k is the wave number given by $k = \mu v/\hbar$. As a result, we can rewrite (2.33) as follows:

$$\Gamma_l = \frac{3\hbar^2\theta_l^2}{\mu R^2} P_l' = 1.25 \times 10^2 \frac{\theta_l^2}{AR^2} P_l' \quad \text{MeV},$$

where $\mu = AM_u$ and $P_l' = kRP_l$. The decay width Γ_γ by way of γ-ray emission is typically on the order of 1 eV.

With use of (2.32), the S-factor is given by

$$S(E) = \frac{656.6}{A} \frac{\omega\Gamma_a\Gamma_b}{(E-E_r)^2 + (\Gamma/2)^2} \exp(bE^{-1/2}) \quad \text{keV barns}.$$

Then (2.27) is integrated to be

$$\langle \sigma v \rangle = \left(\frac{2\pi}{\mu k_B T}\right)^{3/2} \hbar^2 \sum_r S_r \exp\left(-E_r/k_B T\right), \tag{2.34}$$

where $S_r = (\omega\Gamma_a\Gamma_b)/\Gamma$ and the summation is performed over the dominant resonances.

Finally, we obtain the rate of the resonant reaction

$$N_{\mathrm{A}}\langle\sigma v\rangle = 1.54\times10^{5}(AT_9)^{-3/2} \times\sum_r S_r \exp\left(-11.6045E_r/T_9\right) \quad \mathrm{cm}^3\mathrm{mole}^{-1}\mathrm{s}^{-1}, \tag{2.35}$$

where S_r is measured in units of eV barns and E_r is in MeV.

2.4.4 *Inverse Reactions*

In the cosmological environment for relevant particles to react with each other, we may consider that an ordinary reaction and its inverse are in thermal equilibrium, the so-called detailed balance. For the two-body reaction $1+2 \longleftrightarrow 3+4+Q$, the ratio of the reaction rates is written as

$$\frac{\langle\sigma v\rangle_{34}}{\langle\sigma v\rangle_{12}} = f_1\left(\frac{G_1G_2}{G_3G_4}\right)\left(\frac{A_1A_2}{A_3A_4}\right)^{3/2}\exp\left[-\frac{Q}{k_{\mathrm{B}}T}\right]. \tag{2.36}$$

Here Q is the reaction Q-value and the statistical weight f_1 is

$$f_1 = \frac{(2j_1+1)(2j_2+1)(1+\delta_{34})}{(2j_3+1)(2j_4+1)(1+\delta_{12})},$$

where j_i is the spin of the ground state of nucleus i and

$$\delta_{ij} = \begin{cases} 1 & i \mathrm{and} j \mathrm{areidentical}, \\ 0 & \mathrm{otherwise}. \end{cases}$$

The normalized partition function G_i is given as

$$G_i = \frac{1}{2j_i+1}\sum_\mu (2j_i^\mu+1)\exp\left[-\frac{\epsilon_i^\mu}{k_{\mathrm{B}}T}\right],$$

where ϵ_i^μ is the excitation energy of the excited state μ with the spin j_i^μ.

When the particle "4" is a photon, the reverse reaction is photodisintegration of nucleus "3" by thermal photons: $3+\gamma \longrightarrow 1+2$. The reaction rate is

$$\lambda_\gamma = f_2\left(\frac{G_1G_2}{G_3}\right)\left(\frac{A_1A_2}{A_3}\frac{k_{\mathrm{B}}T}{2\pi N_A\hbar^2}\right)^{3/2}\langle\sigma v\rangle_{12}\exp\left[-\frac{Q}{k_{\mathrm{B}}T}\right],$$

which is written as

$$\lambda_\gamma = 9.8686 \times 10^9 T_9^{3/2} f_2 \left(\frac{G_1 G_2}{G_3}\right) \left(\frac{A_1 A_2}{A_3}\right)^{3/2} \times \exp\left[-\frac{11.6045 Q}{T_9}\right] \lambda_{12} \quad \mathrm{s}^{-1}, \tag{2.37}$$

where $\lambda_{12} = N_A \langle \sigma v \rangle_{12}$, Q is measured in units of MeV, and the statistical weight f_2 is given by

$$f_2 = \frac{(2j_1 + 1)(2j_2 + 1)}{(2j_3 + 1)(1 + \delta_{12})}.$$

2.4.5 Rates of Electron Capture and β-Decay

Let ρ be the mass density and μ_e the mean molecular weight of electrons. If we set $\bar{\rho} = \rho/\mu_e$, the electron capture rate at T is given by [7, 8]

$$\lambda_{ec}(\bar{\rho}, T) = -\int_{m_e}^{\infty} \lambda(\varepsilon, 0) \frac{df(\varepsilon, T)}{d\varepsilon} d\varepsilon, \tag{2.38}$$

where m_e is the electron mass and $f(\varepsilon, T)$ is the Fermi-Dirac distribution function:

$$f(\varepsilon, T) = \frac{1}{\exp[(\varepsilon - \varepsilon_{\mathrm{F}})/k_{\mathrm{B}} T] + 1}.$$

The electron Fermi energy ε_{F} is related to $\bar{\rho}$ as

$$\bar{\rho} = 2.1898 \times 10^7 \int_{m_e}^{\infty} \varepsilon(\varepsilon^2 - m_e^2)^{1/2} f(\varepsilon, T) d\varepsilon \quad \mathrm{g\ cm}^{-3},$$

where energy is measured in units of MeV.

The function $\lambda(\varepsilon, 0)$ in (2.38), being the capture rate at $T = 0$ when ε would be regarded as the Fermi energy, is expressed as

$$\lambda(\varepsilon, 0) = \frac{\ln 2}{ft} \int_{W_{\min}}^{W_{\max}} \Lambda(W) dW \tag{2.39}$$

with

$$\Lambda(W) = W(W^2 - 1)^{1/2} \left(\frac{Q_\beta}{m_e} + 1 - W\right)^2 F(Z, W), \tag{2.40}$$

$$W_{\min} = \varepsilon/m_e,$$
$$W_{\max} = Q_\beta/m_e + 1.$$

Here ft is so-called the ft-value of the ordinary β-decay or electron capture, and $\mid Q_\beta \mid$ is the absolute Q-value of the reaction. The atomic number changes from Z to $Z+1$ for β^--decay. The ft-value, $(ft)^-$, obtained from experiments is modified as

$$ft = (ft)^- \frac{(2J+1)_{Z+1}}{(2J+1)_Z},$$

where $(2J+1)_Z$ and $(2J+1)_{Z+1}$ are the spin weights of the mother and daughter nuclei, respectively. The Fermi function $F(Z, W)$ is given by

$$F(Z, W) = 2(1+s)(2pR)^{2(s-1)} \exp(\pi\eta) \left| \frac{\Gamma(s+i\eta)}{\Gamma(2s+1)} \right|^2,$$

where $p = (W^2-1)^{1/2}$, $s = (1-\alpha^2 Z^2)^{1/2}$, $\eta = \alpha Z W/p$, α is the fine structure constant, R is the nuclear radius, and Γ is the gamma function.

The β-decay rate at T is written as

$$\lambda_\beta(\bar{\rho}, T) = \int_{m_e}^{m_e+Q_\beta} \Lambda(W_{\max})[1 - f(\varepsilon, T)]d\varepsilon,$$

where $\Lambda(W_{\max})$ is given by (2.40) with $Z+1$ of the daughter nucleus.

2.4.6 Reaction Rates and Their Uncertainties

We show in Fig. 2.2 the nuclear reaction network relevant to BBN from neutron and proton up to ^{7}Li. Twelve reactions are the most effective to produce light elements.

It is noted that we need to include the *triple* α reaction, 3 ^{4}He $\longrightarrow$ ^{12}C + γ, which could bridge the mass gaps at atomic numbers $A = 5$ and 8 and operate to synthesize further heavy elements in high-density phases of an inhomogeneous universe. Simulations of BBN in the inhomogeneous universe have been performed [9]. See also Ref. [10].

Figures 2.3 and 2.4 show the rates of nuclear reactions, as functions of temperature, from #2 to #12 listed in Fig. 2.2. The sources of the rates are Ref. [11] for p(n, γ)D, Ref. [12] for ^{3}He(n, p)^{3}H and ^{7}Be(n, p)^{7}Li, and Ref. [6] for the others. Uncertainties contained in the rates are rather small for $T_9 < 2$ within 1σ confidence level, which are just the line-widths of the curves in the individual panels.

We take the neutron lifetime $\tau_n = 880.1 \pm 1.1$ s [13].

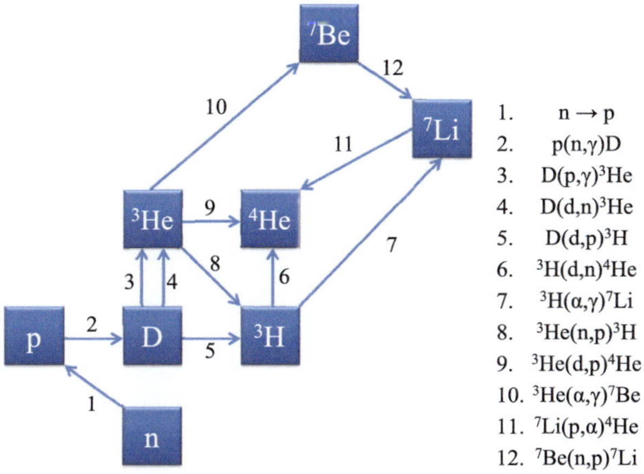

Fig. 2.2 Nuclear reaction network for BBN

2.5 Standard Big Bang Nucleosynthesis

2.5.1 *Overview of SBBN*

Big Bang nucleosynthesis (BBN) offers the deepest reliable probe to the early universe down to nearly 1 s. In an early epoch, $a < 10^{-5}$ radiation term is the most dominant in the Friedmann equation (2.21). The radiation energy density is given by

$$\rho_r = a_{\mathrm{B}} T^4,$$

where a_{B} is the radiation density constant. Then (2.21) reduces to

$$\left(\frac{\dot{a}}{a}\right)^2 = \left(\frac{\dot{T}}{T}\right)^2 = \frac{8\pi G a_{\mathrm{B}}}{3} T^4.$$

When we take the conditions $\dot{T} < 0$ and $T \to \infty$ as $t \to 0$, we obtain a simple solution

$$T = \left(\frac{3}{32\pi G a_{\mathrm{B}}}\right)^{1/4} t^{-1/2}. \tag{2.41}$$

At high temperatures $T_9 > 10$, i.e., $t < 1$ s, neutrons and protons are in equilibrium via weak interactions

$$\mathrm{n} + e^+ \longleftrightarrow \mathrm{p} + \bar{\nu}_e,$$

$$\mathrm{n} + \nu_e \longleftrightarrow \mathrm{p} + e^-,$$

$$\mathrm{n} \longleftrightarrow \mathrm{p} + e^- + \bar{\nu}_e.$$

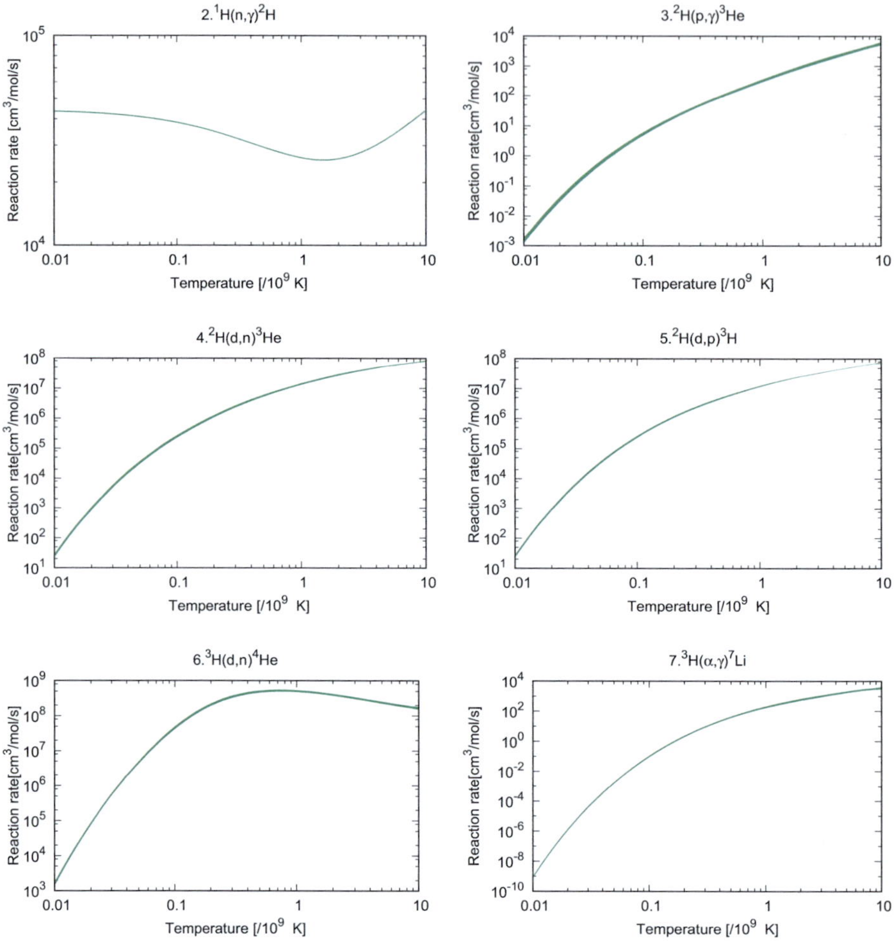

Fig. 2.3 Reaction rates for BBN (Part 1). The line-widths of the curves indicate the uncertainties with 1σ confidence level of the rates

Introducing the chemical potential of nuclear species i as

$$\mu_i = k_{\rm B} T \ln \left[\frac{n_i}{g_i} \frac{h^3}{(2\pi m_i k_{\rm B} T)^{3/2}} \right], \tag{2.42}$$

where h is the Planck constant and n_i, g_i, and m_i are the number density, statistical weight, and mass of nuclide i, and then the chemical equilibrium

$$\mu_n + m_n = \mu_p + m_p$$

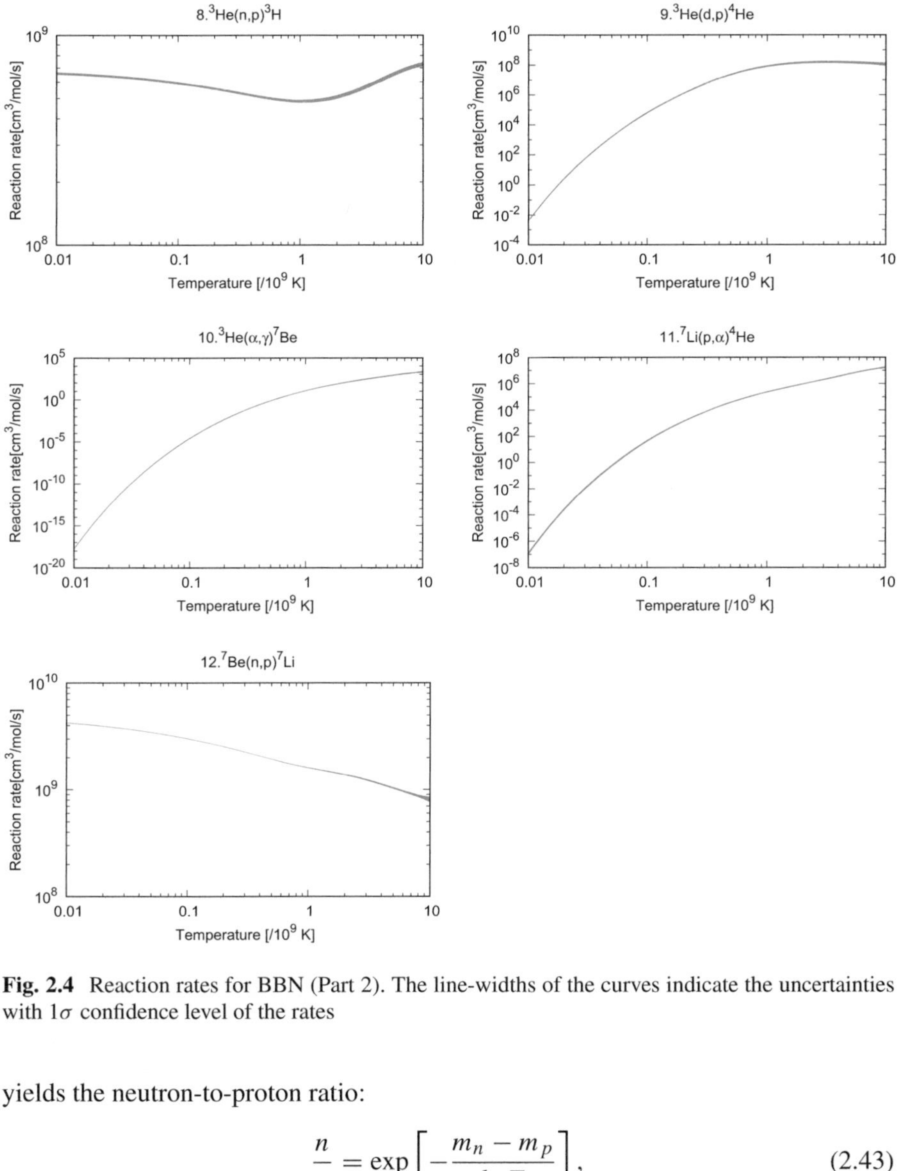

Fig. 2.4 Reaction rates for BBN (Part 2). The line-widths of the curves indicate the uncertainties with 1σ confidence level of the rates

yields the neutron-to-proton ratio:

$$\frac{n}{p} = \exp\left[-\frac{m_n - m_p}{k_B T}\right], \tag{2.43}$$

where we set conventionally $n = n_n$ and $p = n_p$.

When T_9 is 12, i.e., $t = 1.5$ s, electron-positron pairs annihilate. Consequently, β-equilibrium cannot be sustained, but neutrons begin to decay freely with the lifetime τ_n. The neutron fraction is represented by

$$X_n(t) = \frac{n}{n+p} = X_n(t_\beta)\exp\left[-\frac{t-1.5}{\tau_n}\right], \tag{2.44}$$

where $X_n(t_\beta) = 0.23$ is evaluated from (2.43) at $t = 1.5$ s.

During the stage $T_9 > 2$ the following reaction is in equilibrium:

$$\mathrm{p} + \mathrm{n} \longleftrightarrow \mathrm{D} + \gamma.$$

Then we obtain

$$\mu_n + m_n + \mu_p + m_p = \mu_D + m_D.$$

Using (2.42) with $g_n = g_p = 2$ and $g_D = 3$, we have

$$\frac{n_n n_p}{n_D} = \frac{4}{3}\frac{(\pi m_p k_\mathrm{B} T)^{3/2}}{h^3}\exp\left[-\frac{m_n + m_p - m_D}{k_\mathrm{B} T}\right]. \tag{2.45}$$

It is noted that abundance of D increases rapidly as temperature decreases. When $T_9 = 0.8$, we get $n_n \simeq n_D$ at $t = 350$ s. Hence (2.44) gives $X_n \simeq 0.15$.

Subsequently, nuclear reactions proceed to build up light elements. Figure 2.2 shows the nuclear reaction network which is relevant to BBN. Since there exist no stable nuclei with mass numbers A = 5 and 8, almost all neutrons of $X_n \simeq 0.15$ fuse with protons to produce ^{4}He. The mass fraction of ^{4}He amounts to $Y = 2X_n \simeq 0.3$, which is roughly consistent with the observed abundance of ^{4}He.

2.5.2 Numerical Calculations of SBBN

Our standard BBN (hereafter SBBN) model is constructed in the framework of the following items:

1. Dynamics is described by the Friedmann equation (2.21) with a constant Λ term.
2. Radiation is the most dominant in the early universe.
3. Neutrinos are three species (ν_e, ν_μ and ν_τ) and non-degenerate.

In the early universe $T_9 > 10$, photons, neutrinos, and electron-positron pairs are in thermal equilibrium. The energy density is given by

$$\rho_r = \rho_\gamma + \rho_\nu + \rho_e.$$

Neutrinos decouple from other particles at $T_9 \simeq 20$. Then pair annihilation of electrons and positrons converts their energy into photons. We can follow photon temperature as described in Appendix B.

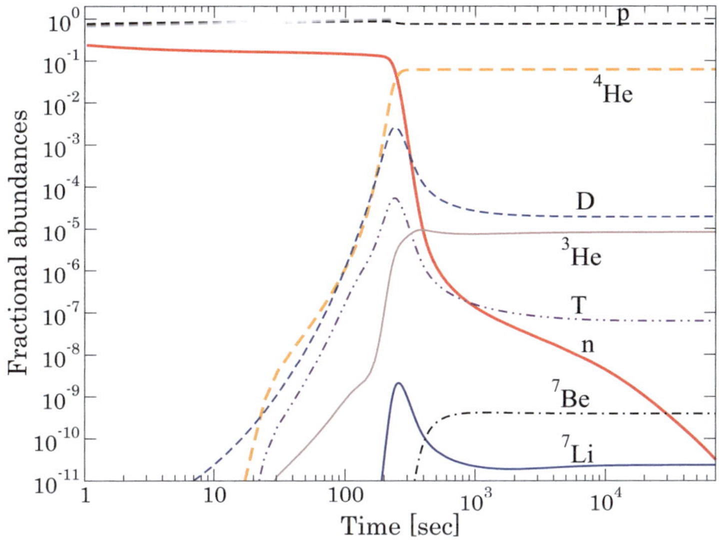

Fig. 2.5 Evolution of abundance during BBN era for $\eta_{10} = 6.19$

SBBN contains only one free parameter, the baryon-to-photon ratio $\eta = n_b/n_\gamma$, which is related to the baryon density parameter as

$$\Omega_b = 3.66 \times 10^{-3}\eta_{10}h^{-2}, \tag{2.46}$$

where $\eta_{10} = 10^{10}\eta$.

Once we specify a value of η_{10}, then we obtain a numerical solution $x(t)$ of (2.23) and hence evolution of ρ_b and T. Using the nuclear reaction network, we perform numerical calculations for the evolution of the abundance of chemical elements as described in Appendix C.

Figure 2.5 shows the evolution of light elements during the BBN stage for $\eta_{10} = 6.19$. Note that once a significant amount of D is produced at $t \geq 30\,\mathrm{s}$, nuclear reactions proceed further to synthesize ^{4}He. It can be seen that the abundance is frozen out at $t \simeq 10^3$ s except for β-decay of neutrons. The products are mainly ^{4}He and a trace of D, T, ^{3}He, ^{7}Be, and ^{7}Li.

Figure 2.6 shows the produced abundance as a function of η_{10}. The upper panel is the mass fraction Y of ^{4}He, and the middle and lower are the number ratios D/H and ^{7}Li/H, respectively. The line-widths of the curves indicate uncertainties contained in the nuclear reaction rates. It is noted that Y is insensitive to η, while D depends strongly and monotonically on η. This is the reason that D is called a baryometer. In a low-density range, $\eta_{10} < 2$, ^{7}Li is produced directly through $^3\mathrm{H} + {}^4\mathrm{He} \longrightarrow {}^7\mathrm{Li} + \gamma$, while at $\eta_{10} > 3$, it is originated from electron capture or β^+-decay (half-life $\tau_{1/2} = 53.22$ days) of ^{7}Be produced via $^3\mathrm{He} + {}^4\mathrm{He} \longrightarrow {}^7\mathrm{Be} + \gamma$.

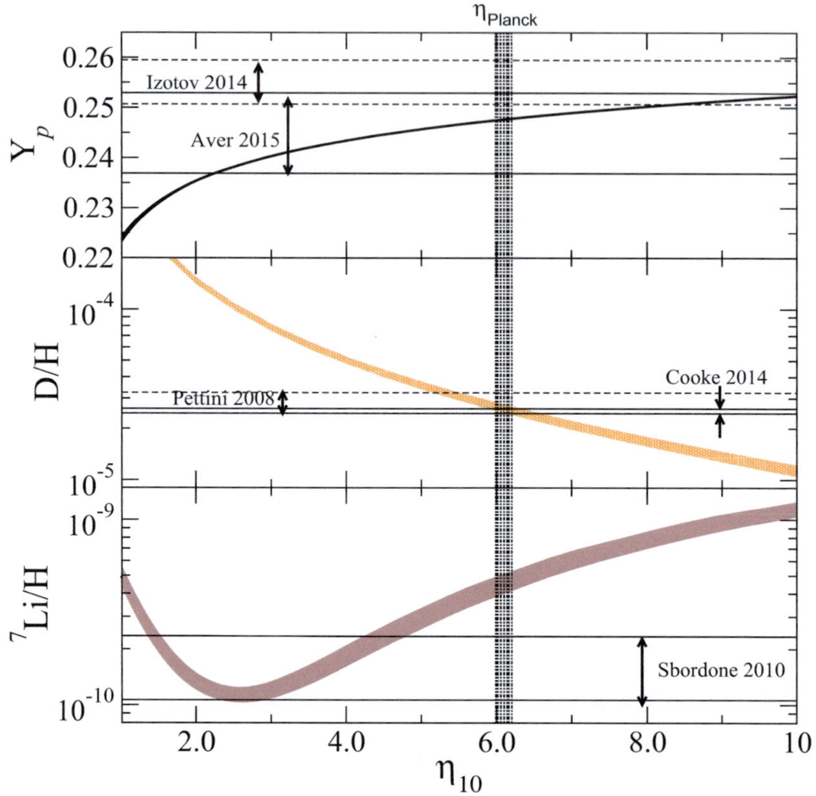

Fig. 2.6 Abundance of light elements produced in SBBN. The horizontal lines indicate the observed primordial abundances [14–18]. (The vertical shaded region is taken from *Planck* [3])

2.5.3 *Observed Abundance of Light Elements*

As has been stated repeatedly, the produced amount of ^{4}He is sensitive to the neutron-to-proton ratio and hence the physical circumstances at the onset of nucleosynthesis. On the other hand, the abundance of D strongly depends on the baryon density. BBN fails to produce heavy elements. They are synthesized in massive stars during the evolution of galaxies. Therefore, the abundance of heavy elements, i.e., metals, is a measure of age. The primordial abundance is deduced from observations of objects with low metallicity. There exist very large spreads in some observed abundances of light elements due to different observational methods. Let us describe how we adopt the observed primordial abundance.

Helium abundance is derived from observations of He I (neutral helium) emission lines from low-metallicity blue compact dwarf galaxies. The observed helium abundance Y^* is obtained from He I $\lambda 10830$ Å emission line of 45 H II (ionized hydrogen) regions in 43 dwarf galaxies (1 Å = 10^{-8} cm). Using a linear regression of Y^* versus O/H, the oxygen-to-hydrogen ratio, the primordial abundance Y_p of ^{4}He is determined to be [14]

$$Y_\mathrm{p} = 0.2551 \pm 0.0022. \tag{2.47}$$

On the other hand, using a Markov chain Monte Carlo analysis for 15 qualifying objects, an alternative low value is reported [15]:

$$Y_\mathrm{p} = 0.2449 \pm 0.0040. \tag{2.48}$$

It is considered that the significant differences between (2.47) and (2.48) are ascribable to the differences in determining Y^* for individual objects and in selecting the samples.

Deuterium abundance is derived from D and H absorption lines toward high-redshift quasars due to intervening intergalactic clouds. The primordial D abundance is determined from measures of a very metal-poor damped Lyman α system (DLA) at redshift $z = 2.618$ to be [16]

$$\mathrm{D/H} = (2.82 \pm 0.19) \times 10^{-5}. \tag{2.49}$$

Moreover, a precise D abundance is reported from measurements of DLA at $z = 3.067$ [17]:

$$\mathrm{D/H} = (2.53 \pm 0.04) \times 10^{-5}. \tag{2.50}$$

Lithium abundance is derived from Li I absorption lines in metal-poor dwarf stars, because Li is easily destroyed through ^{7}Li(p, α)^{4}He at temperatures around 2×10^6 K in stellar interiors. The primordial abundance of ^{7}Li is determined from measurements of Li I λ6798 Å absorption line of 28 dwarfs in the galactic halo to be [18]

$$^7\mathrm{Li/H} = (1.58 \pm 0.31) \times 10^{-10}. \tag{2.51}$$

It is noted that a high primordial abundance

$$^7\mathrm{Li/H} = (2.75 - 4.17) \times 10^{-10},$$

is proposed [19] by taking account of significant depletion or destruction during the lifetime of dwarf stars. Note also that some fraction of ^{7}Li and particularly all of ^{6}Li are produced together with Be and B through spallation of CNO nuclei by cosmic ray protons and α particles.

We show in Fig. 2.6 the observed abundances Y_p's of (2.47) and (2.48), D/H's of (2.49) and (2.50), and ^{7}Li/H of (2.51) by the horizontal lines with the identifications Izotov 2014 [14], Aver 2015 [15], Pettini 2008 [16], Cooke 2014 [17], and Sbordone 2010 [18], respectively. The vertical shaded region indicates the range $\eta_{10} = 6.07 \pm 0.06$ determined from *Planck* [3]. Note that the produced amounts of ^{4}He and D in SBBN agree well with the observed values, though unfortunately ^{7}Li provides poor consistency. It seems that the standard model is not enough to explain the discrepancy in the abundance of ^{7}Li, which encourages us to research nonstandard models.

2.5.4 Neutron Lifetime

The update of reaction rates does not remove discrepancies between the results of SBBN and the observed abundances. Now we examine the neutron lifetime τ_n, because it is one of the most important items of physical input which affects the produced amount of ^{4}He. We take $\tau_n = 885.7$ s [20], which was frequently used in the previous evaluations, and 880.1 s [21] that is the currently recommended value. Table 2.3 gives the consistent range of η_{10} through the produced abundance of SBBN for two cases of neutron lifetime.

We show in Fig. 2.7 a record of measured lifetimes of neutrons in recent years. The data are compiled in [24]. It is surprising that the evaluated lifetime depends significantly on the methods. Measurements have been pursued by using two methods: proton-counting and neutron-counting methods [25]. The results exhibit a large discrepancy for the neutron lifetime.

Table 2.3 Range of η_{10} for different neutron lifetimes

	η_{10}		
	$\tau_n = 880.1$ s	$\tau_n = 885.7$ s	Reference
D	6.08 − 6.54	6.09 − 6.56	[17]
Y_p	≥6.77	≥6.00	[22]
Y_p	≥2.33	≥4.48	[23]

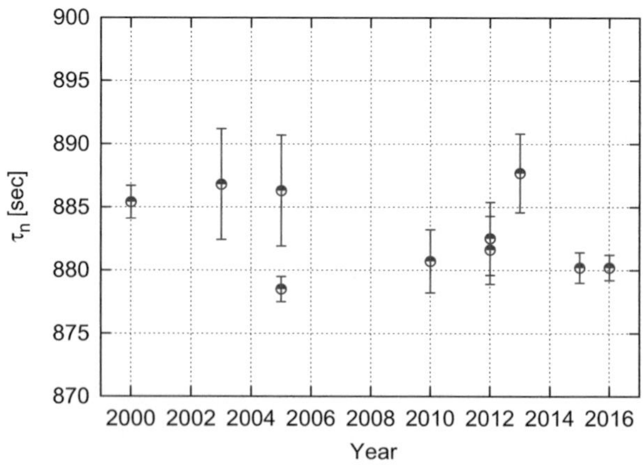

Fig. 2.7 Neutron lifetimes measured in recent years. (The data are taken from [24])

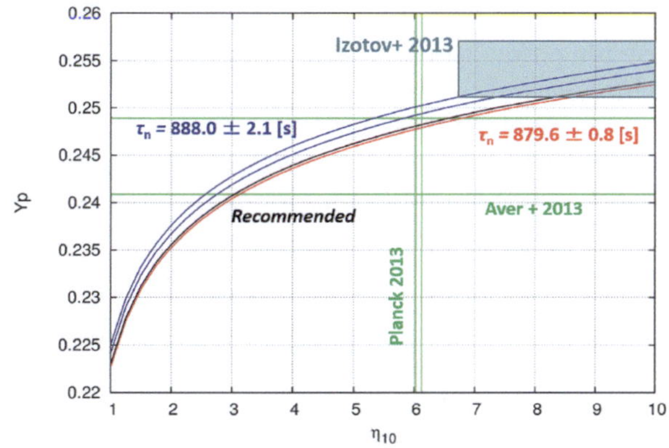

Fig. 2.8 Dependence of the produced helium abundance on the neutron lifetime. (The observed abundances are from [22] and [23])

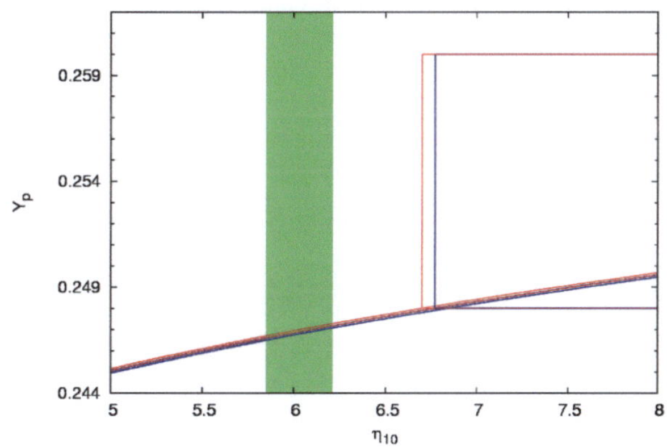

Fig. 2.9 Dependence of the produced helium abundance on the nuclear reaction rates. The red and blue lines represent the results calculated with the rates of [6] and [12], respectively. The boxes indicate the ranges of η_{10} constrained from the observed abundance [22]. (The shaded vertical band is the constraint from *Planck* [3])

Figure 2.8 shows the produced amount of ^{4}He, using alternative lifetimes of neutrons. The difference is small. When we adopt the recommended value of τ_n and the observed abundance of Aver 2013 [23], the permitted range of η_{10} is barely consistent with that from *Planck*. However, SBBN fails to reconcile the higher abundance of Izotov 2013 [22]. We show in Fig. 2.9 the produced abundance of ^{4}He, using the nuclear reaction rates given in Ref. [6] or [12]. The boxes represent the ranges of η_{10} constrained from the observed abundance [22], which are inconsistent with the vertical band constraint from *Planck* [3].

2.6 Another Observational Success of the Standard Model

As stated in the previous section, one of the most outstanding successes of the standard Big Bang model is the quantitative prediction of primordial abundance for light elements. The amounts of D, ^{4}He, and ^{7}Li produced during *the first three minutes* are in good agreement with the observed abundance.

Now we consider the magnitude-redshift relation of distant objects as a potential test for cosmological models, because measures of the cosmological distance depend sensitively on the spatial curvature and the expansion dynamics.

Since light emitted from a distant object is redshifted due to the expansion of the universe, redshift is used to describe the distance to the emitting source. If emitted light of wavelength λ is observed by a local observer at wavelength λ_0, the wavelength increases proportionally to the scale factor a. Then the redshift z is defined by

$$1 + z = \frac{\lambda_0}{\lambda} = \frac{a_0}{a}. \tag{2.52}$$

A source with luminosity L_S at a distance d in our Galaxy is observed with energy flux $F = L_S/(4\pi d^2)$. Generalizing this to the expanding universe, we define the luminosity distance

$$d_L = \left(\frac{L_S}{4\pi F}\right)^{1/2}. \tag{2.53}$$

The energy of light emitted from the object with time interval Δt is denoted as ΔE, whereas the observer receives the energy ΔE_0 with the time interval Δt_0. The luminosities L_S and L_0 are given by

$$L_S = \frac{\Delta E}{\Delta t}, \qquad L_0 = \frac{\Delta E_0}{\Delta t_0}. \tag{2.54}$$

Since the photon energy is proportional to its frequency, we have

$$\frac{\Delta E}{\Delta E_0} = \frac{\nu}{\nu_0} = \frac{\Delta t_0}{\Delta t} = \frac{\lambda_0}{\lambda} = 1 + z. \tag{2.55}$$

From (2.54) and (2.55), we obtain

$$L_S = L_0\,(1+z)^2\,. \tag{2.56}$$

We find from the metric (2.14) that the area of the sphere of the comoving coordinate r at $t = t_0$ is $S = 4\pi a_0^2 r^2$. Hence the observed energy flux is

$$F = \frac{L_0}{4\pi a_0^2 r^2}. \tag{2.57}$$

Substituting (2.56) and (2.57) to (2.53), we obtain the luminosity distance

$$d_L = (1+z) a_0 r. \tag{2.58}$$

For the emitted light which propagates along the radial direction toward the origin, we can set $ds = d\theta = d\varphi = 0$ in (2.14). We then obtain

$$\chi = \int_0^r \frac{dr}{(1-kr^2)^{1/2}} = \int_t^{t_0} \frac{dt}{a(t)}. \tag{2.59}$$

The first integral yields

$$\chi = \begin{cases} \sin^{-1} r & (k=+1), \\ r & (k=0), \\ \sinh^{-1} r & (k=-1). \end{cases}$$

Using (2.52), the second integral becomes

$$\chi = \int_0^z \frac{dz}{H},$$

where H is represented from (2.23) as

$$H = H_0[\Omega_r(1+z)^4 + \Omega_m(1+z)^3 + \Omega_k(1+z)^2 + \Omega_\Lambda]^{1/2}. \tag{2.60}$$

The apparent magnitude m of the source at the luminosity distance d_L is related to the absolute magnitude M through

$$m = M + 5\log\frac{d_L}{10\ \mathrm{pc}}, \tag{2.61}$$

(1 pc $= 3.0857 \times 10^{18}$ cm). It is convenient to use the distance modulus defined as

$$\mu_d = m - M = 5\log d_L + 25, \tag{2.62}$$

where d_L is measured in units of Mpc. As can be seen from (2.58), d_L is a function of z. Hence (2.62) is called the magnitude-redshift $(m-z)$ relation.

Type Ia supernovae (SNe Ia) are ideal standard candles, because they are found in all kinds of galaxies and easily identified due to a characteristic shape of light curves and spectral features that follow a standard pattern of evolution. They are so luminous that can be observed in distant galaxies beyond $z \simeq 1$. There is significantly less scatter in the maximum luminosities. The absolute maximum magnitude is $M = -19.6 \pm 0.2$ at B band [26], which is comparable to the absolute magnitude $\simeq -20.3$ of our Galaxy. SNe Ia are considered to be explosions of white dwarfs that accrete matter from binary companions and just reach the Chandrasekhar limit, the maximum possible mass of $\simeq 1.4\,M_\odot$ that can be supported by degenerate pressure of electrons ($1\,M_\odot = 1.99 \times 10^{33}$ g). It follows that the nature of explosion does not depend much on the environment in the universe.

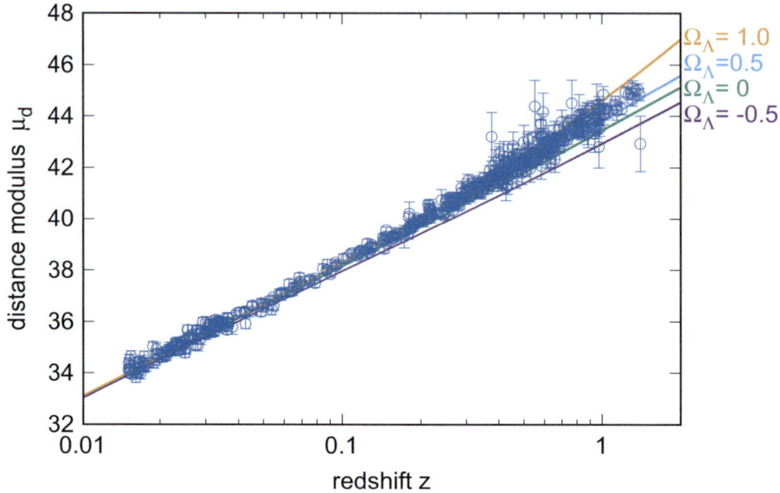

Fig. 2.10 Magnitude-redshift relation of SNe Ia [27]. Theoretical curves correspond to flat models with $\Omega_\Lambda = -0.5$, 0, 0.5, and 1.0

Figure 2.10 shows the $m - z$ relation of SNe Ia, where the data are taken from the Union2 compilation [27] up to $z = 1.12$. Theoretical curves are also shown for flat models with $\Omega_\Lambda = -0.5$, 0, 0.5, and 1.0. In a model with a large positive Ω_Λ, the expansion is accelerated, and hence the distance modulus becomes large.

We examine χ^2 analysis (see Sect. 3.1.3 for χ^2 analysis in detail) for the distance moduli of SNe Ia:

$$\chi^2(\Omega_m, \Omega_\Lambda) = \sum_i \frac{|\mu_d^i(z; \Omega_m, \Omega_\Lambda) - \mu_{\rm obs}^i(z)|^2}{(\sigma_{\rm obs}^i)^2 + (\sigma_v^i)^2}, \tag{2.63}$$

where $\mu_{\rm obs}^i$ and $\sigma_{\rm obs}^i$ are the observed distance modulus and its uncertainty, respectively, σ_v^i is the dispersion in the redshift due to the peculiar velocity v given by

$$\sigma_v^i = \left(v \frac{d\mu_d^i}{dz} \right).$$

We adopt $v = 300\,\mathrm{km\,s^{-1}}$, i.e., 10^{-3}, in units of $c = 1$. The total number of the samples is $N = 557$ in the Union2 compilation [27].

Figure 2.11 shows the contours with 1σ (68.3%), 2σ (95.4%), and 3σ (99.7%) confidence levels on the $(\Omega_m, \Omega_\Lambda)$ plane. The flat model ($k = 0$) is indicated by the dashed line with $\Omega_m + \Omega_\Lambda = 1$. It is suggested that cosmological models without dark energy are excluded even if we adopt the non-zero curvature term.

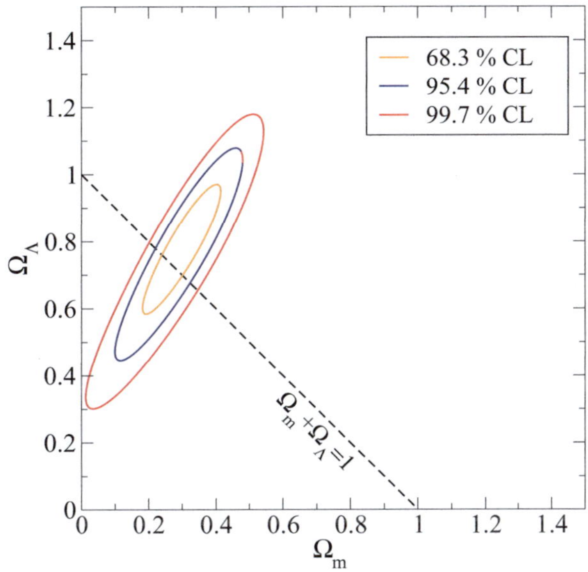

Fig. 2.11 Confidence regions in the $(\Omega_m, \Omega_\Lambda)$ plane for SNe Ia. The observed data are from Union2 compilation [27]. The dashed line indicates the flat universe $(\Omega_m + \Omega_\Lambda = 1)$

Table 2.4 Cosmological parameters from $m - z$ relation of SNe Ia

Description	Symbol	$\Omega_k \neq 0$	$\Omega_k = 0$
Matter density	Ω_m	0.30 ± 0.08	0.27 ± 0.02
Dark energy density	Ω_Λ	0.79 ± 0.13	0.72 ± 0.02

The best fit values of Ω_m and Ω_Λ are given in Table 2.4. If we adopt the flat universe suggested by the observations of CMB anisotropy [2, 3], these parameters are consistent with the results of *Planck* (Table. 2.1).

References

1. Weinberg, S.: Cosmology, p. 41. Oxford University Press, New York (2008)
2. Hinshaw, G., et al.: Nine-year Wilkinson microwave anisotropy probe (WMAP) observations: cosmological parameter results. Astrophys. J. Suppl. **208**, 19 (2013). https://doi.org/10.1088/0067-0049/208/2/19
3. Planck Collaboration: Planck 2015 results. XIII. Cosmological parameters. Astron. Astrophys. **594**, A13 (2016). https://doi.org/10.1051/0004-6361/201525830
4. Trimble, V.: Existence and nature of dark matter in the universe. Ann. Rev. Astron. Astrophys. **25**, 425–472 (1987). https://doi.org/10.1146/annurev.aa.25.090187.002233
5. Clayton, D.D.: Principles of Stellar Evolution and Nucleosynthesis, p. 283. McGraw-Hill, New York (1968)
6. Xu, Y., et al.: NACRE II: an update of the NACRE compilation of charged-particle-induced thermonuclear reaction rates for nuclei with mass number $A < 16$. Nucl. Phys. A **918**, 61–169 (2013). https://doi.org/10.1016/j.nuclphysa.2013.09.007
7. Egawa, Y., Yokoi, K., Yamada, M.: Electron capture in highly evolved stars. Prog. Theor. Phys. **54**, 1339–1355 (1975). https://doi.org/10.1143/PTP.54.1339

8. Egawa, Y., Yokoi, K.: Energy loss accompanying the electron captures in highly evolved stars. Prog. Theor. Phys. **57**, 1255–1261 (1975). https://doi.org/10.1143/PTP.57.1255
9. Nakamura, R., Hashimoto, M., Fujimoto, S., Sato, K.: Constraint on heavy element production in inhomogeneous big-bang nucleosynthesis from the light element observations. J. Astrophys. **2013**, 587294 (2013). https://doi.org/10.1155/2013/587294
10. Nakamura, R., Hashimoto, M., Ichimasa, R., Arai, K.: Big bang nucleosynthesis: constraints on nuclear reaction rates, neutrino degeneracy, inhomogeneous and Brans-Dicke models. Int. J. Mod. Phys. E **26**, 1741003 (2017). https://doi.org/10.1142/S0218301317410038
11. Ando, S., Cyburt, R.H., Hong, S.W., Hyun, C.H.: Radiative neutron capture on a proton at big-bang nucleosynthesis energies. Phys. Rev. C **74**, 025809 (2006). https://doi.org/10.1103/PhysRevC.74.025809
12. Descouvemont, P., Adahchour, A., Angulo, C., Coc, A., Vangioni-Flam, E.: Compilation and R-matrix analysis of big bang nuclear reaction rates. Atomic Data Nucl. Data Tables. **88**, 203–236 (2004). https://doi.org/10.1016/j.adt.2004.08.001
13. Beringer, J., et al.: Review of particle physics. Phys. Rev. D **86**, 010001 (2012). https://doi.org/101103/PhysRevD.86.010001
14. Izotov, Y.I., Thuan, T.X., Guseva, N.G.: A new determination of the primordial He abundance using the He I λ 10830 A emission line: cosmological implications. Mon. Not. R. Astron. Soc. **445**, 778–793 (2014). https://doi.org/10.1093/mnras/stu1771
15. Aver, E., Olive, K.A., Skillman, E.D.: The effects of He I λ10830 on helium abundance determinations. J. Cosmol. Astropart. Phys. **07**, 011 (2015). https://doi.org/10.1088/1475-7516/2015/07/011
16. Pettini, M., et al.: Deuterium abundance in the most metal-poor damped Lyman alpha system: converging on $\Omega_{b,0}h^2$. Mon. Not. R. Astron. Soc. **391**, 1499–1510 (2008). https://doi.org/10.1111/j.1365-2966.2008.13921.x
17. Cooke, R.J., et al.: Precision measures of the primordial abundance of deuterium. Astrophys. J. **781**, 31 (2014). https://doi.org/10.1088/0004-647X/781/1/31
18. Sbordone, L., et al.: The metal-poor end of the spite plateau. I. Stellar parameters, metallicities, and lithium abundances. Astron. Astrophys. **522**, A26 (2010). https://doi.org/10.1051/0004-6361/200913282
19. Korn, A., et al.: A probable stellar solution to the cosmological lithium discrepancy. Nature **442**, 657–659 (2006). https://doi.org/10.1038/nature05011
20. Serebrov, A., et al.: Measurement of the neutron lifetime using a gravitational trap and a low-temperature fomblin coating. Phys. Lett. B **605**, 72–78 (2005). https://doi.org/10.1016/j.physletb.2004.11.013
21. Beringer, J., et al.: Review of particle physics. Phys. Rev. D **86**, 010001 (2012). https://doi.org/10.1103/PhysRevD.86.010001
22. Izotov, Y.I., Stasińska, G., Guseva, N.G.: Primordial ^{4}He abundance: a determination based on the largest sample of H II regions with a methodology tested on model H II regions. Astron. Astrophys. **558**, A57 (2013). https://doi.org/10.1051/0004-6361/201220782
23. Aver, E., et al.: The primordial helium abundance from updated emissivities. J. Cosm. Astropart. Phys. **1311**, 071 (2013). https://doi.org/10.1088/1475-7516/2013/11/017
24. Patringnahi, C., et al.: (Particle data group): review of particle physics. Chin. Phys. C **40** 100001 (2016). https://doi.org/10.1088/1674-1137/40/10/100001
25. Dubbers, D., Schmidt, M.G.: The neutron and its role in Cosmology and particle physics. Rev. Mod. Phys. **83** 1111–1171 (2011). https://doi.org/10.1103/RevModPhys.83.1111
26. Branch, D., Tammann, G.A.: Type Ia supernovae as standard candles. Ann. Rev. Astron. Astrophys. **30**, 359–389 (1992). https://doi.org/10.1146/annurev.aa.30.090192.002043
27. Amanullah, R., et al.: (The supernova cosmology project): spectra and HST light curves of six type Ia supernovae at $0.511 < z < 1.12$ and the Union2 compilation. Astrophys. J. **716**, 712 (2010). https://doi.org/10.1088/0004-637X/716/1/712

Chapter 3
Theories Beyond the Standard Model

Abstract Many alternative models are proposed, though remarkable agreement is obtained on the primordial abundance of standard BBN. As an approach to nonstandard models, first we investigate a possibility that neutrinos or antineutrinos are degenerate. An excess density of neutrinos speeds up the expansion of the universe, leaving more neutrons at the onset of nucleosynthesis. In addition, degenerate electron-neutrinos shift less neutrons through β-equilibrium. Performing χ^2 analysis for the calculated and observed abundances of ^{4}He and D, we determine the plausible ranges of the degeneracy parameter and the baryon-to-photon ratio. Next, we explore BBN under the brane-world cosmology. The Friedmann-like equation is derived in the five-dimensional universe. It is found that more ^{4}He is produced than in standard BBN because of rapid expansion due to the interaction energy on the brane. Finally, we examine thermal evolution in the early universe including a decaying cosmological term which is treated as a source of the gravitational field.

Keywords Brane world · Decaying cosmological term · Five-dimensional space-time · Neutrino degeneracy

3.1 Lepton Asymmetry

Baryon asymmetry exists in our universe: baryons slightly dominate over antibaryons. Since pairs of baryon and antibaryon annihilate into photons at the epoch $kT_{\rm B} \simeq 1\,{\rm GeV}$, we obtain $n_b + n_{\bar{b}} \simeq n_\gamma$, where $n_{\bar{b}}$ is the number density of antibaryons. Then the baryon asymmetry $(n_b - n_{\bar{b}})/(n_b + n_{\bar{b}})$ in the early era $k_{\rm B}T > 1\,{\rm GeV}$ is nearly equal to the baryon to photon ratio $(n_b/n_\gamma)_0 \simeq 10^{-9}$ at the present epoch.

On the other hand, there is a possibility that leptons have large asymmetry [1]. As stated in Sect. 2.5 of SBBN, the produced abundance of ^{4}He is sensitive to the neutron-to-proton ratio at the onset of nucleosynthesis. If neutrinos are degenerate, the energy density increases, which leads to speed up in the expansion rate, leaving more neutrons, and eventually enhancing the produced amount of ^{4}He. Moreover, if electron-neutrinos degenerate, β-equilibrium shifts to less neutrons and hence

M. Hashimoto et al., *Big-Bang Nucleosynthesis*, SpringerBriefs in Physics,
https://doi.org/10.1007/978-981-13-2935-7_3

lowers the primordial abundance of ^{4}He. Therefore, it is worthwhile to investigate effects of neutrino degeneracy on the produced abundance of BBN.

3.1.1 Reaction Rates Between Neutrons and Protons

The number density of neutrinos (or anti-neutrinos) with momentum between p and $p + dp$ at temperature T_ν is given by the Fermi distribution

$$n_\nu(p)dp = \frac{4\pi p^2}{h^3} dp \left[\exp\left(\frac{p \pm \mu_\nu}{k_B T_\nu}\right) + 1\right]^{-1}, \tag{3.1}$$

where μ_ν is the chemical potential of neutrinos; the minus sign refers to neutrinos and the plus to antineutrinos. The net neutrino number density is

$$n_\nu = \int_0^\infty dp[n_\nu(p) - n_{\bar{\nu}}(p)]. \tag{3.2}$$

If we set $y = p/k_B T_\nu$ and $x = \mu_\nu/k_B T_\nu$, (3.2) is written as

$$n_\nu = 4\pi \left(\frac{k_B T_\nu}{h}\right)^3 \mathcal{N}(x), \tag{3.3}$$

where

$$\mathcal{N}(x) = \int_0^\infty y^2 dy \left[\frac{1}{\exp(y - x) + 1} - \frac{1}{\exp(y + x) + 1}\right]. \tag{3.4}$$

Since both $n_\nu a^3$ and $T_\nu a$ are constant during the expansion of the universe, the quantity n_ν / T_ν^3 and hence $\mathcal{N}(x)$ are also kept constant.

It is beneficial to introduce the degeneracy parameter

$$\xi_i = \frac{\mu_{\nu_i}}{k_B T_\nu}, \tag{3.5}$$

where the subscript i stands for electron-, muon-, and tau-neutrinos. Note that ξ_i is preserved as the universe expands. Performing integration of (3.4), we obtain the degree of lepton asymmetry

$$\eta_\nu = \frac{n_\nu}{n_\gamma} = \frac{1}{12\zeta(3)} \left(\frac{T_\nu}{T}\right)^3 (\pi^2 \xi_i + \xi_i{}^3), \tag{3.6}$$

where the number density of photons n_γ is given by (2.24).

Although it is pointed out that the chemical potentials of three generations of neutrino are the same due to neutrino oscillation, we take the case where only the electron-neutrinos are degenerate, and we set $\xi_\mu = \xi_\tau = 0$. With the use of (3.1), the energy density of neutrinos can be written as

$$\rho_\nu = \frac{4\pi}{h^3}\int_0^\infty dE \frac{E^3}{\exp[(E \pm \mu_\nu)/k_{\rm B}T_\nu] + 1}. \tag{3.7}$$

Then the energy density of three generations of neutrinos and antineutrinos is

$$\rho_\nu + \rho_{\bar{\nu}} = a_{\rm B}{T_\nu}^4 F_i(\xi_i), \tag{3.8}$$

where $a_{\rm B}$ is the radiation density constant and

$$F_i(\xi_i) = \frac{21}{8} + \frac{15}{4}\left(\frac{\xi_i}{\pi}\right)^2 + \frac{15}{8}\left(\frac{\xi_i}{\pi}\right)^4.$$

When $k_{\rm B}T > 2\,\text{MeV}$, photons, electrons, positrons, and neutrinos are in thermal equilibrium ($T = T_\nu$). Then the total energy density of radiation is

$$\rho_r = \left[1 + \frac{7}{4} + F_i(\xi_i)\right] a_{\rm B}T^4. \tag{3.9}$$

It is noted that neutrinos decouple from the other particles at $k_{\rm B}T \simeq 1.5\,\text{MeV}$ and neutrino temperature varies as $T_\nu \sim a^{-1}$. After pair annihilation of electrons and positrons, as described in Appendix B, photon temperature rises up to $T = (11/4)^{1/3}T_\nu$. Radiation is composed of photons and neutrinos, so that we have

$$\rho_r = \left[1 + \left(\frac{4}{11}\right)^{4/3} F_i(\xi_i)\right] a_{\rm B}T^4. \tag{3.10}$$

Figure 3.1 shows the neutrino temperature in the BBN stage. It can be seen that degenerate neutrinos increase the expansion rate and hence lower the temperature than the standard model.

When electron-neutrinos are degenerate, reaction rates for n $\longleftrightarrow$ p are significantly altered. Particularly, the neutron-to-proton ratio in equilibrium is given by

$$\frac{n}{p} = \exp\left[-\frac{m_n - m_p}{k_{\rm B}T} - \xi_e\right]. \tag{3.11}$$

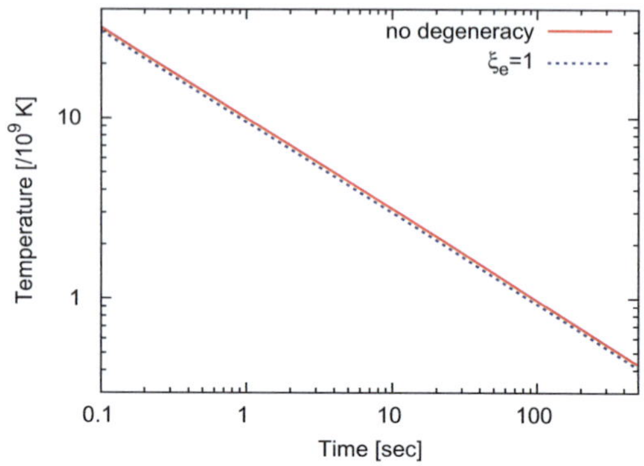

Fig. 3.1 Neutrino temperature in the BBN stage (Red: $\xi_e = 0$, Blue: $|\xi_e| = 1.0$)

The reaction rates are written as follows [2]:

$$\lambda_{\mathrm{n}\to\mathrm{pe}\nu} = \frac{1}{\tau_n\lambda_0}\int_1^q d\epsilon \frac{\epsilon(\epsilon-q)^2(\epsilon^2-1)^{1/2}}{[1+\exp(-\epsilon u)][1+\exp((\epsilon-q)u_\nu+\xi_e)]}, \quad (3.12)$$

$$\lambda_{\mathrm{n}\nu\to\mathrm{pe}} = \frac{1}{\tau_n\lambda_0}\int_q^\infty d\epsilon \frac{\epsilon(\epsilon-q)^2(\epsilon^2-1)^{1/2}}{[1+\exp(-\epsilon u)][1+\exp((\epsilon-q)u_\nu-\xi_e)]}, \quad (3.13)$$

$$\lambda_{\mathrm{ne}\to\mathrm{p}\nu} = \frac{1}{\tau_n\lambda_0}\int_1^\infty d\epsilon \frac{\epsilon(\epsilon+q)^2(\epsilon^2-1)^{1/2}}{[1+\exp(\epsilon u)][1+\exp(-(\epsilon+q)u_\nu+\xi_e)]}, \quad (3.14)$$

$$\lambda_{\mathrm{pe}\nu\to\mathrm{n}} = \frac{1}{\tau_n\lambda_0}\int_1^q d\epsilon \frac{\epsilon(\epsilon-q)^2(\epsilon^2-1)^{1/2}}{[1+\exp(\epsilon u)][1+\exp((q-\epsilon)u_\nu+\xi_e)]}, \quad (3.15)$$

$$\lambda_{\mathrm{pe}\to\mathrm{n}\nu} = \frac{1}{\tau_n\lambda_0}\int_q^\infty d\epsilon \frac{\epsilon(\epsilon-q)^2(\epsilon^2-1)^{1/2}}{[1+\exp(\epsilon u)][1+\exp((q-\epsilon)u_\nu-\xi_e)]}, \quad (3.16)$$

$$\lambda_{\mathrm{p}\nu\to\mathrm{ne}} = \frac{1}{\tau_n\lambda_0}\int_1^\infty d\epsilon \frac{\epsilon(\epsilon+q)^2(\epsilon^2-1)^{1/2}}{[1+\exp(-\epsilon u)][1+\exp((q+\epsilon)u_\nu+\xi_e)]}, \quad (3.17)$$

where τ_n is the lifetime of neutrons, $u = m_e/k_{\mathrm{B}}T$, $u_\nu = m_e/k_{\mathrm{B}}T_\nu$, $\epsilon = E_e/m_e$, $q = (m_n - m_p)/m_e$, and

$$\lambda_0 = \int_1^q d\epsilon\epsilon(\epsilon-q)^2(\epsilon^2-1)^{1/2} = 1.63609.$$

We show in Fig. 3.2 the rates (3.12), (3.13), (3.14), (3.15), (3.16), and (3.17) for $\xi_e = -0.046$ normalized to those of SBBN. It is noted that neutrons and protons are in β-equilibrium at temperatures $T_9 > 20$. The crucial reactions in SBBN are $\mathrm{p}+e^- \longrightarrow \mathrm{n}+\nu$ and $\mathrm{n}+e^+ \longrightarrow \mathrm{p}+\bar{\nu}$ in the range $T_9 \simeq 2-20$. When $\xi_e = -0.046$, the reactions $\mathrm{p}+\bar{\nu} \longrightarrow \mathrm{n}+e^+$ and $\mathrm{n}+\nu \longrightarrow \mathrm{p}+e^-$ are dominated.

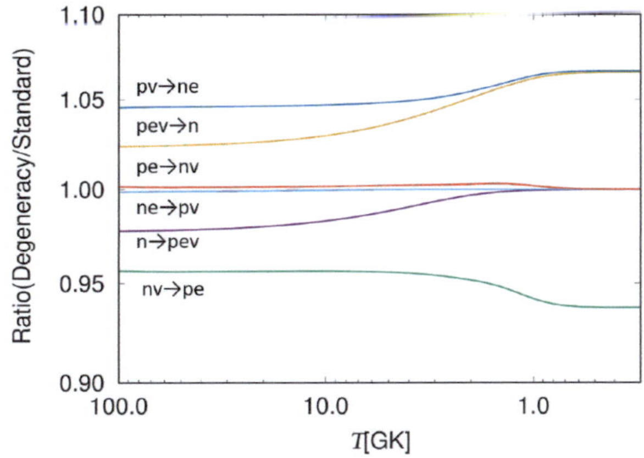

Fig. 3.2 Rates of n $\longleftrightarrow$ p reactions for $\xi_e = -0.046$ normalized to those of SBBN ($\xi_e = 0$)

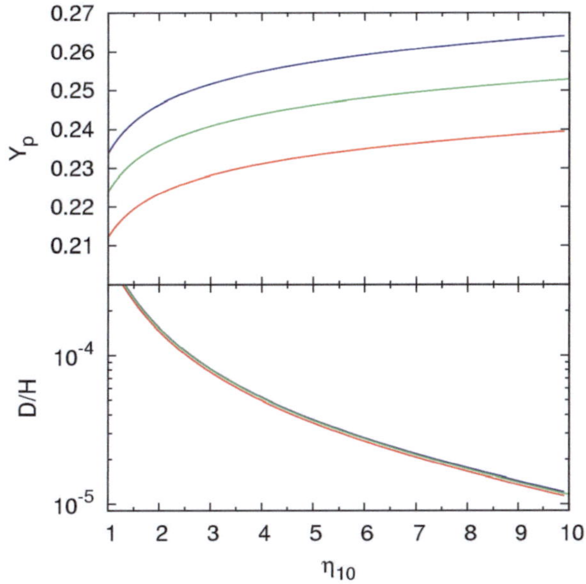

Fig. 3.3 Produced abundance of Y_p and D/H in a neutrino degenerate model (blue, $\xi_e = -0.1$; green, $\xi_e = 0.0$; red, $\xi_e = 0.1$)

3.1.2 BBN with Degenerate Neutrinos

Using the initial n/p ratio (3.11) at high temperature for a specified value of ξ_e, we perform BBN calculations. The numerical technique is explained in Appendix C. Figure 3.3 shows the produced abundances for $\xi_e = -0.1$, 0, and 0.1 as a function of the baryon-to-photon ratio η_{10}. It can be seen that the amount of ^{4}He strongly depends on ξ_e. In the case of positive ξ_e, the rates for n $\longrightarrow$ p are enhanced, so that there remain less neutrons, which yields less ^{4}He than in SBBN.

On the contrary, the produced amount of D is nearly the same as that in SBBN. The resultant abundance is compared with the primordial abundances of $Y_{\rm p}$ and D/H inferred from observations [3, 4], as described in Sect. 2.5.3.

3.1.3 χ^2 Analysis

To find consistent ranges of η and ξ_e between the calculated and observed abundances, we carry out χ^2 analysis. Let us introduce the statistical quantity

$$\chi^2(\eta, \xi_e) = \sum_i \frac{|Y^i_{\rm th}(\eta, \xi_e) - Y^i_{\rm obs}|^2}{(\sigma^i_{\rm th})^2 + (\sigma^i_{\rm obs})^2}. \tag{3.18}$$

Here $Y^i_{\rm th}$ and $Y^i_{\rm obs}$ are the theoretical (calculated) and observed values, respectively, $\sigma^i_{\rm th}$ and $\sigma^i_{\rm obs}$ are their standard deviations from the average values, and the summation i is taken over $Y_{\rm p}$ and D/H. Note that $\sigma_{\rm th}$ is obtained from 1σ errors associated with relevant nuclear reaction rates.

Figure 3.4 shows the contours with 1σ, 2σ, and 3σ confidence levels (CL) on the (η_{10}, ξ_e) plane. We find that the consistent region is

$$-0.04 \le \xi_e \le -0.01, \qquad 6.1 \le \eta_{10} \le 6.5$$

within 1σ CL.

Now we explain a relation between confidence levels and the number of parameters [5]. To measure the reliabilitywe consider

$$\Delta\chi^2 = \chi^2 - \chi^2_{\rm min}. \tag{3.19}$$

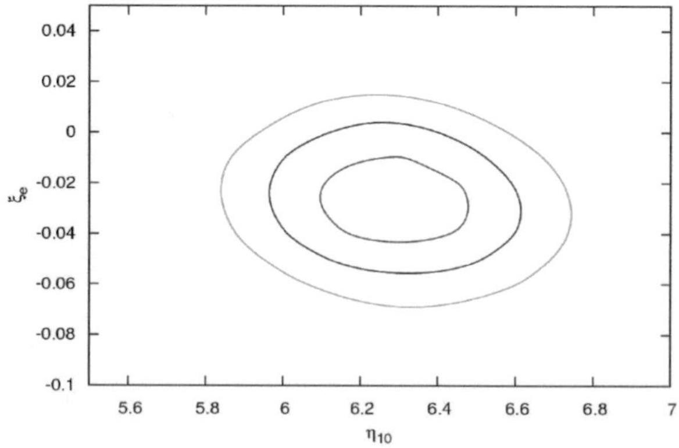

Fig. 3.4 Contours with 1σ, 2σ, and 3σ confidence levels from $Y_{\rm p}$ and D/H

Table 3.1 $\Delta\chi^2$ given from confidence levels and the number of parameters

CL	1	2	3	4	5
68.3 %	1.00	2.30	3.53	4.72	5.89
95.4 %	2.71	4.61	6.25	7.78	9.24
99.7 %	9.00	11.8	14.2	16.3	18.2

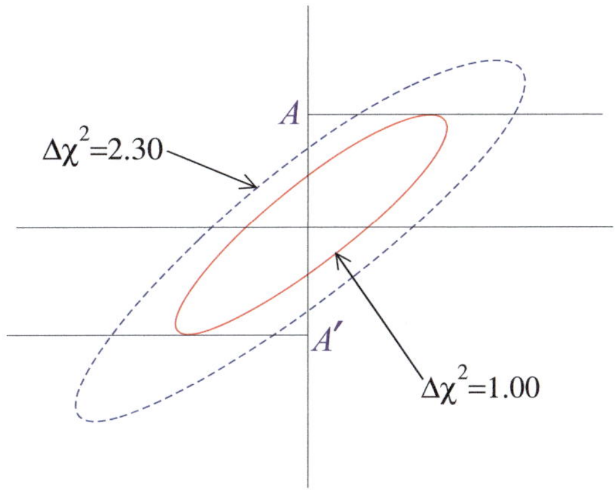

Fig. 3.5 Confidence regions which indicate values of χ^2 larger than the fitted minimum $\chi^2_{\min}$. The dashed curve indicates $\Delta\chi^2 = 2.30$, and the ellipse contains 68.3% normally distributed data of two parameters. The solid curve with $\Delta\chi^2 = 1.00$ is projected onto the vertical axis with the interval AA' for one parameter. Other numerical values are given in Table 3.1

Note that the lowest value $\chi^2_{\min}$ indicates the highest reliability. Table 3.1 gives $\Delta\chi^2$. The first column is CL and the first line indicates the number of parameters. For a set of two parameters, reliability of $\Delta\chi^2 = 2.30$ is 68.3% and that of $\Delta\chi^2 = 4.61$ is 95.4%.

An example of contours obtained from χ^2 analysis is shown in Fig. 3.5, where the two axes correspond to individual parameters. Two parameters distribute inside the contour of $\Delta\chi^2 = 2.30$ with 1σ (68.3%) CL. When we project a contour onto the axis, we obtain $\Delta\chi^2$ for one parameter. For example, the contour of $\Delta\chi^2 = 1.00$, which corresponds to 1σ CL for one parameter (see Table 3.1), is mapped on the vertical axis as shown in Fig. 3.5. Consequently, we confirm the range from A to A' with 1σ CL.

3.2 BBN Under the Brane Cosmology

3.2.1 Five-Dimensional Einstein Tensor

A brane cosmology has been invented in the frame of superstring theory (more details are given in several reviews, such as [6]). Our four-dimensional universe is considered to be a brane on the surface of a five-dimensional space-time [7, 8] which is called "bulk" as shown in Fig. 3.6.

Based on the method [9, 10], we derive the five-dimensional Friedmann-like equation from Einstein's equation

$$G_{AB} = R_{AB} - \frac{1}{2} g_{AB} R = \kappa^2_{(5)} T_{AB}, \tag{3.20}$$

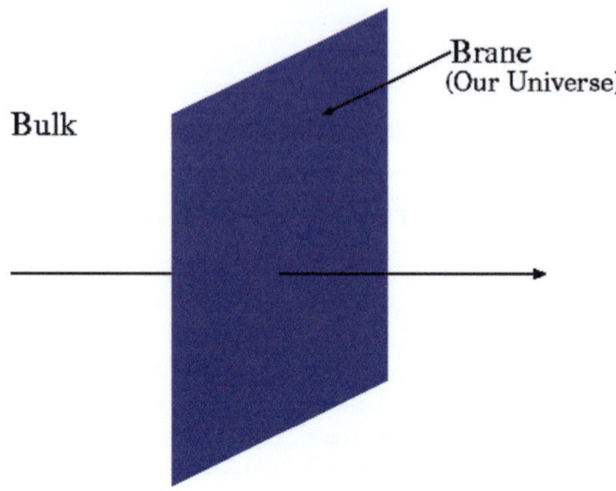

Fig. 3.6 Schematic feature of a brane universe

where the suffixes A and B run 0, 1, ... 4 and $\kappa^2_{(5)}$ corresponds to the five-dimensional Newton's gravitational constant. We can introduce five-dimensional Planck mass $M_{(5)}$ which represents the energy scale of the grand unified theory such that $M^3_{(5)} = 1/(2\kappa_{(5)})^2$.

The line element of our brane world in a five-dimensional space-time is given by

$$\begin{aligned} ds^2_{(5)} &= g_{AB} dx^A dx^B \\ &= -n^2(\tau, y)dt^2 + a^2(\tau, y)\gamma_{ij} dx^i dx^j + b^2(\tau, y)dy^2. \end{aligned} \tag{3.21}$$

Here the extra dimension is expressed by the direction y and

$$\gamma_{ij} = X^{-2}\delta_{ij} \tag{3.22}$$

with

$$X = 1 + \frac{k}{4}\delta_{mn}x^m x^n,$$

where the roman suffix runs 1, 2, and 3 and k is the curvature constant.

The nonvanishing Christoffel symbols (5.2) are written as follows:

$$\Gamma^0_{00} = \frac{\dot{n}}{n}, \qquad \Gamma^0_{ij} = \frac{\dot{a}a}{n^2}\gamma_{ij}, \qquad \Gamma^0_{05} = \frac{n'}{n},$$

$$\Gamma^0_{55} = \frac{\dot{b}b}{n^2}, \qquad \Gamma^i_{0j} = \frac{\dot{a}}{a}\delta^i{}_j,$$

$$\Gamma^i_{jk} = -\frac{k}{2}X^{-1}x^m(\delta_{jm}\delta^i{}_k + \delta_{km}\delta^i{}_j - \delta^i{}_m\delta_{jk}),$$

$$\Gamma^i{}_{j5} = \frac{a'}{a}\delta^i{}_j, \qquad \Gamma^5_{00} = \frac{n'n}{b^2}, \qquad \Gamma^5_{05} = \frac{\dot{b}}{b},$$

$$\Gamma^5_{ij} = -\frac{a'a}{b^2}\gamma_{ij}, \qquad \Gamma^5_{55} = \frac{b'}{b},$$

where the prime denotes the differentiation with respect to y.

The Ricci tensors (5.8) are

$$R_{00} = -3\frac{\ddot{a}}{a} - \frac{\ddot{b}}{b} + \frac{n''n}{b^2} - \frac{b'n'n}{b^3} + 3\frac{\dot{a}\dot{n}}{an} + 3\frac{a'n'n}{ab^2} + \frac{\dot{b}\dot{n}}{bn},$$

$$R_{ij} = \gamma_{ij}\Big(\frac{\ddot{a}a}{n^2} + 2\frac{\dot{a}^2}{n^2} - \frac{\dot{n}\dot{a}a}{n^3} - \frac{a''a}{b^2} - 2\frac{a'^2}{b^2} + \frac{\dot{a}\dot{b}a}{bn^2} - \frac{n'a'a}{b^2n}$$
$$+\frac{n'a'a}{nb^2} + \frac{b'a'a}{b^3} + 2k\Big),$$

$$R_{05} = 3\left(-\frac{\dot{a}'}{a} + \frac{\dot{a}n'}{an} + \frac{a'\dot{b}}{ab}\right),$$

$$R_{55} = -\frac{n''}{n} - 3\frac{a''}{a} + \frac{\ddot{b}b}{n^2} - \frac{\dot{n}\dot{b}b}{n^3} + 3\frac{\dot{a}\dot{b}b}{an^2} + \frac{b'n'}{bn} + 3\frac{a'b'}{ab}.$$

Then the curvature scalar (5.9) is

$$R = 6\frac{\ddot{a}}{an^2} + 2\frac{\ddot{b}}{bn^2} - 2\frac{n''}{nb^2} + 2\frac{b'n'}{nb^3} - 6\frac{\dot{a}\dot{n}}{an^3} - 6\frac{a'n'}{ab^2n} - 2\frac{\dot{b}\dot{n}}{bn^3}$$
$$+6\frac{\dot{a}^2}{a^2n^2} - 6\frac{a'^2}{a^2b^2} - 6\frac{a''}{ab^2} + 6\frac{\dot{a}\dot{b}}{abn^2} + 6\frac{a'b'}{ab^3} + 6\frac{k}{a^2}.$$

We obtain the nonvanishing Einstein tensors (5.10)

$$G_{00} = 3\left[\frac{\dot{a}}{a}\left(\frac{\dot{a}}{a} + \frac{\dot{b}}{b}\right) - \frac{n^2}{b^2}\left\{\frac{a''}{a} + \frac{a'}{a}\left(\frac{a'}{a} - \frac{b'}{b}\right)\right\} + \frac{kn^2}{a^2}\right], \tag{3.23}$$

$$G_{ij} = \gamma_{ij}\left(-2\frac{\ddot{a}a}{n^2} - \frac{\dot{a}^2}{n^2} + 2\frac{\dot{n}\dot{a}a}{n^3} + \frac{a'^2}{b^2} + 2\frac{a''a}{b^2} - 2\frac{\dot{b}\dot{a}a}{bn^2} + 2\frac{n'a'a}{b^2n}\right.$$
$$\left.-2\frac{b'a'a}{b^3} - \frac{\ddot{b}a^2}{bn^2} + \frac{n''a^2}{nb^2} - \frac{n'b'a^2}{nb^3} + \frac{\dot{b}\dot{n}a^2}{bn^3} - k\right), \tag{3.24}$$

$$G_{05} = 3\left(-\frac{\dot{a}'}{a} + \frac{\dot{a}n'}{an} + \frac{a'\dot{b}}{ab}\right), \tag{3.25}$$

$$G_{55} = 3\left[\frac{a'}{a}\left(\frac{a'}{a} + \frac{n'}{n}\right) - \frac{b^2}{n^2}\left\{\frac{\ddot{a}}{a} + \frac{\dot{a}}{a}\left(\frac{\dot{a}}{a} - \frac{\dot{n}}{n}\right)\right\} - \frac{kb^2}{a^2}\right]. \tag{3.26}$$

3.2.2 Friedmann-Like Equation

The energy-momentum tensor is decomposed into the parts of brane and bulk:

$$T^A{}_B = T^A{}_B|_{\text{bulk}} + T^A{}_B|_{\text{brane}}. \tag{3.27}$$

Here $T^A{}_B|_{\text{bulk}}$ and $T^A{}_B|_{\text{brane}}$ are defined by

$$T^A{}_B|_{\text{bulk}} = \text{diag}(-\rho_B, P_B, P_B, P_B, P_T), \tag{3.28}$$

$$T^A{}_B|_{\text{brane}} = \frac{\delta(y)}{b}\text{diag}(-\rho_{br}, P_{br}, P_{br}, P_{br}, 0), \tag{3.29}$$

where $T^A{}_B|_{\text{brane}}$ is the energy-momentum tensor on the brane at $y = 0$. Since the constituent on the brane is assumed to be homogeneous and isotropic, ρ_B, P_B, P_T, ρ_{br}, and P_{br} depend only on t. Furthermore, we assume

$$-\rho_B = P_B = P_T. \tag{3.30}$$

The (0,0) component of Einstein's equation (3.20) is from (3.23), (3.28), and (3.29) written as

$$3\left[\frac{\dot{a}}{a}\left(\frac{\dot{a}}{a}+\frac{\dot{b}}{b}\right)-\frac{n^2}{b^2}\left\{\frac{a''}{a}+\frac{a'}{a}\left(\frac{a'}{a}-\frac{b'}{b}\right)\right\}+\frac{kn^2}{a^2}\right]$$
$$= \kappa_{(5)}^2 n^2\left(\frac{\delta(y)}{b}\rho_{br}+\rho_B\right). \tag{3.31}$$

When we consider the situation on the brane, we need a connection condition between bulk and brane at $y = 0$. We integrate (3.31) in a narrow range $-\varepsilon \le y \le \varepsilon$, ($\varepsilon > 0, \varepsilon \to 0$). Consequently, only the term with a'' contributes in the left hand side and the first term remains in the right hand side due to the δ-function. Therefore, we obtain

$$\int_{-\varepsilon}^{\varepsilon} a''dy = a'_{0+} - a'_{0-} = -\frac{\kappa_{(5)}^2}{3}a_0 b_0 \rho_b, \tag{3.32}$$

where a_0 and b_0 are evaluated at $y = 0$ and the suffix $0+$ indicates the limiting case from $y > 0$ to 0 and $0-$ is that from $y < 0$ to 0. We assume the boundary condition

$$a(\tau, y_0) = a(\tau, -y_0).$$

Then we have

$$a'(\tau, y_0) = -a'(\tau, -y_0),$$

It follows that

$$a'_{0+} - a'_{0-} = 2a'_{0+} = -2a'_{0-}.$$

Therefore, from (3.32) we get

$$a_0'^2 = \frac{\kappa_{(5)}^4}{36} a_0^2 b_0^2 \rho_b^2. \tag{3.33}$$

Let us introduce a function

$$F(\tau, y) = \frac{(a'a)^2}{b^2} - \frac{(\dot{a}a)^2}{n^2} - ka^2. \tag{3.34}$$

Differentiating (3.34) with respect to y, we get

$$F' = \frac{2a'a^3}{b^2}\left[\frac{a''}{a} + \frac{a'}{a}\left(\frac{a'}{a} - \frac{b'}{b}\right)\right] - \frac{2\dot{a}a^3}{n^2}\left[\frac{\dot{a}'}{a} + \frac{\dot{a}}{a}\left(\frac{a'}{a} - \frac{n'}{n}\right)\right] - 2kaa'.$$

From $G_{05} = 0$ we have

$$\frac{\dot{a}'}{a} = \frac{\dot{a}n'}{an} + \frac{a'\dot{b}}{ab}.$$

Therefore, we obtain

$$F' = -\frac{2a'a^3}{3n^2} G_{00}.$$

Similarly, differentiation of (3.34) with respect to t gives

$$\dot{F} = \frac{2\dot{a}a^3}{3b^2} G_{55}.$$

Consequently, Einstein's equation yields

$$F' = \frac{2a'a^3}{3} \kappa_{(5)}^2 T^0{}_0|_{\text{bulk}}, \tag{3.35}$$

$$\dot{F} = \frac{2\dot{a}a^3}{3} \kappa_{(5)}^2 T^5{}_5|_{\text{bulk}}. \tag{3.36}$$

Since $T^0{}_0|_{\text{bulk}} = -\rho_B$ is independent of y, (3.35) is integrated to be

$$F + \frac{\kappa_{(5)}^2}{6} a^4 \rho_B + C = 0, \tag{3.37}$$

where C is a function which depends only on t. On the other hand, from (3.28) and (3.30) we have $T^5{}_5|_{\text{bulk}} = -\rho_B$. Then the derivative of (3.35) with respect to t and the derivative of (3.36) with respect to y lead to $\dot{\rho}_B = 0$. It follows that integration of (3.36) yields the same result as (3.37). Therefore, C becomes constant.

From (3.37) and (3.34), we obtain

$$\left(\frac{\dot{a}}{na}\right)^2 = \frac{1}{6}\kappa_{(5)}^2\rho_B + \left(\frac{a'}{ba}\right)^2 - \frac{k}{a^2} + \frac{C}{a^4}. \tag{3.38}$$

On the brane $y = 0$, we can set $n_0 = 1$. Using (3.33), we get

$$\left(\frac{\dot{a}_0}{a_0}\right)^2 = \frac{1}{6}\kappa_{(5)}^2\rho_B + \frac{\kappa_{(5)}^4}{36}\rho_{br}^2 + \frac{C}{a_0^4} - \frac{k}{a_0^2}. \tag{3.39}$$

The first term in the right hand side corresponds to the incoming energy density from the bulk, the second is the interaction energy density on the brane, the third behaves as a_0^{-4} like the radiation energy density, and the last is the same curvature term as in the Friedmann equation (2.21).

Henceforth, we omit the subscript "0," because we are concerned with quantities on the brane. We divide the energy density into two components

$$\rho_{br} = \Lambda + \rho, \tag{3.40}$$

where Λ is a constant called "brane tension" and ρ is the ordinary energy density. Then we obtain

$$\left(\frac{\dot{a}}{a}\right)^2 = \frac{1}{6}\kappa_{(5)}^2\rho_B + \frac{\kappa_{(5)}^4}{36}\Lambda^2 + \frac{\kappa_{(5)}^4}{18}\Lambda\rho + \frac{\kappa_{(5)}^4}{36}\rho^2 + \frac{C}{a^4} - \frac{k}{a^2}. \tag{3.41}$$

Following Refs. [7, 8], we take a relation

$$\frac{1}{6}\kappa_{(5)}^2\rho_B + \frac{1}{36}\kappa_{(5)}^4\Lambda^2 = 0, \tag{3.42}$$

and we also adopt the following identification: to recover the standard cosmology

$$\frac{1}{6}\kappa_{(5)}^4\Lambda = 8\pi G. \tag{3.43}$$

Consequently, we get

$$\left(\frac{\dot{a}}{a}\right) = \frac{8\pi G}{3}\rho + \frac{C}{a^4} + \frac{1}{36}\kappa_{(5)}^4\rho^2 - \frac{k}{a^2}.$$

Since the second term in the right hand side behaves like the radiation, we set $C/a^4 = 8\pi G\rho_{DR}/3$, where ρ_{DR} is called the "dark radiation" density. Therefore, the Friedmann-like equation on the brane is written as [9–11]

$$H^2 = \left(\frac{\dot{a}}{a}\right)^2 = \frac{8\pi G}{3}(\rho + \rho_{DR}) + \frac{1}{36}\kappa_{(5)}^4\rho^2 - \frac{k}{a^2}. \tag{3.44}$$

3.2.3 BBN on the Brane

In our brane-world cosmology, the expansion rate of the universe is different from that in the standard model, because it contains two extra terms: dark radiation ρ_{DR} and interaction energy on the brane ρ^2. It is noted that these terms affect the primordial nucleosynthesis [12–14]. We examine how the interaction energy ρ^2 influences the yields of BBN.

We assume the flat universe ($k = 0$) and neglect the contribution of ρ_{DR}. Then (3.44) reduces to

$$H^2 = \frac{8\pi G}{3}\rho + \frac{1}{144M_{(5)}^6}\rho^2. \tag{3.45}$$

Figure 3.7 shows the expansion rate H during the BBN era for $M_{(5)} = 10$ and 15 TeV. The rate deviates significantly from that of the standard model in an early stage, where the second term in (3.45) becomes dominant.

We make BBN calculations using the nuclear reaction network. Figure 3.8 shows the abundance of ^{4}He produced in the brane universe for $M_{(5)} = 10$ and 15 TeV. It can be seen that, compared to SBBN indicated with the dashed line, much more ^{4}He is produced. This is because more neutrons survive due to more rapid expansion.

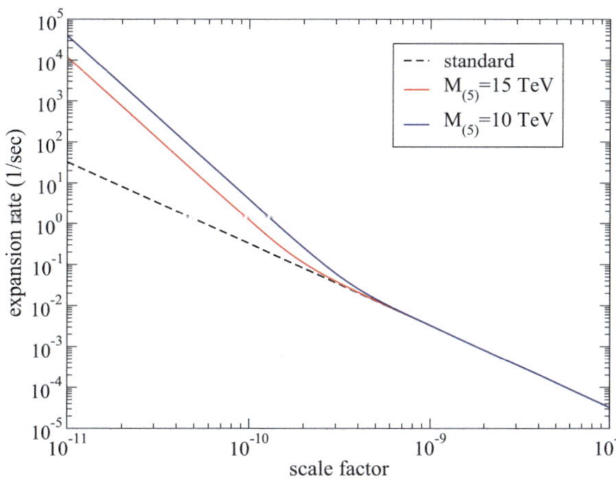

Fig. 3.7 Expansion rate of the brane universe with $M_{(5)} = 10$ and 15 TeV. The dashed line denotes the standard model

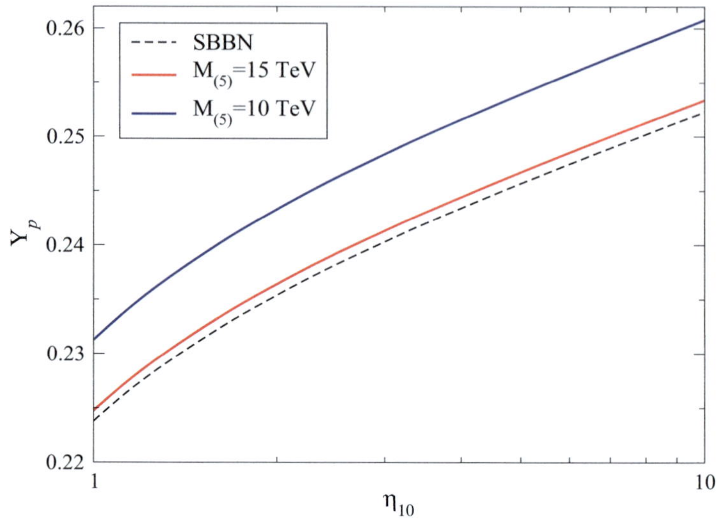

Fig. 3.8 Helium abundance produced in the brane universe. The blue and red lines are $M_{(5)} = 10$ and 15 TeV, respectively. The dashed line corresponds to SBBN

Fig. 3.9 Contours with 1σ and 2σ confidence levels from Y_p

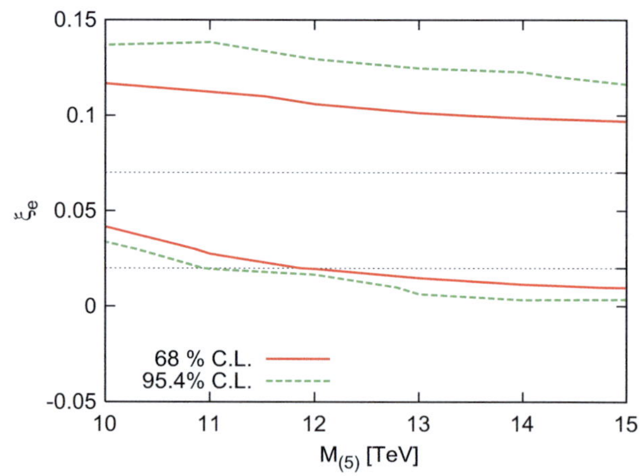

Note that the produced amounts of D and ^{7}Li are almost the same as in SBBN since they are not so sensitive to the neutron-to-proton ratio.

When we compare our results with the observed abundance, as described in Sect. 2.5.3 for SBBN, we can unfortunately find no reasonable range of the baryon-to-photon ratio η. Consequently, we are forced to take the condition $M_{(5)} \to \infty$, which reduces to the standard model.

To reduce the abundance of ^{4}He, we incorporate neutrino degeneracy with a positive ξ_e as explained in Sect. 3.1.2. We perform the χ^2 analysis for two parameters $M_{(5)}$ and ξ_e. Figure 3.9 shows a part of the contours with 1σ (68 %) and 2σ (95 %) CLs on the $(M_{(5)}, \xi_e)$ plane, where $\eta_{10} = 6.1$ is taken to be fixed. It is noted that overproduction of ^{4}He due to inclusion of the interaction energy on the brane is canceled out if electron-neutrinos are degenerate.

3.3 Phenomenological Variable Λ Models

The cosmological term is considered to be extremely large in the early universe, so that it is not constant, but decreasing to a small value around $10^{-56}\,\text{cm}^{-2}$ at the present epoch as inferred from observations [15]. Let us examine thermal evolution of the universe with the decaying cosmological term. The term is transferred to the right hand side in Einstein's field equation (2.2) and treated as a source of the gravitational field [16].

The energy-momentum conservation law including variable Λ is written as

$$\frac{d}{da}\left[\left(\rho+\frac{\Lambda}{8\pi G}\right)a^3\right]=-3\left(p-\frac{\Lambda}{8\pi G}\right)a^2. \tag{3.46}$$

The Friedmann equation (2.21) is

$$H^2=\left(\frac{\dot{a}}{a}\right)^2=\frac{8\pi G}{3}\rho-\frac{k}{a^2}+\frac{\Lambda}{3}. \tag{3.47}$$

We adopt an equation of state: $p=(\gamma-1)\rho$.

When we examine the thermal history of a realistic universe, we take a general relation for the energy density which includes both matter and radiation:

$$\rho=\rho_m+\rho_r. \tag{3.48}$$

Here the radiation energy density contains photons, electron-positron pairs, and neutrinos:

$$\rho_r=\alpha T^4 \tag{3.49}$$

with

$$\alpha=a_{\text{B}}\left(\frac{11}{4}+\frac{7}{8}N_\nu\right),$$

where $N_\nu=3$ is the number of neutrino species.

Neglecting the contribution from matter, pressure is given by

$$p=\frac{1}{3}\alpha T^4. \tag{3.50}$$

When we normalize the scale factor a to its present value, as has been done below (2.20) in Sect. 2.3, (3.46) and (3.47) are rewritten in terms of the density parameters

$$\frac{d\ln a}{dt} = H_0\left[\Omega + \lambda + \frac{\Omega_k}{a^2}\right]^{1/2}, \tag{3.51}$$

$$\frac{d\Omega}{d\ln a} = -4\Omega + \frac{\Omega_{m0}}{a^3} - \frac{d\lambda}{d\ln a}, \tag{3.52}$$

where

$$\Omega = \frac{\rho}{\rho_{\rm cr}}, \qquad \lambda = \frac{\Lambda}{3H_0^2}, \qquad \Omega_k = -\frac{k}{H_0^2}.$$

Note that both Ω and λ are not constants but functions of a.

Now we investigate the thermal evolution of variable Λ models. The Λ term is assumed to be of the form

$$\lambda = \lambda_0 a^{-m}. \tag{3.53}$$

The energy equation (3.52) becomes

$$\frac{d\Omega}{d\ln a} = -4\Omega + \frac{\Omega_{m0}}{a^3} + m\lambda. \tag{3.54}$$

We consider nonsingular models quantitatively. From the conditions $da/dt = d^2a/dt^2 = 0$ at the moment of *bounce*, the critical value λ_* of λ_0 and the minimum value a_* are found to be [16]

$$\lambda_* = \frac{(3\gamma-2)(3\gamma-m)\Omega_0}{3\gamma(2-m)a_*^{3\gamma-m} + (3\gamma-2)m}, \tag{3.55}$$

$$3\gamma(3\gamma-m)\Omega_0 a_*^2 + 3\gamma(2-m)(1-\Omega_0)a_*^{3\gamma} - (3\gamma-2)(3\gamma\Omega_0-m)a_*^m = 0. \tag{3.56}$$

The numerical solutions for various sets of γ, Ω_0, and m are given in Table 3.2. The critical temperature is defined as $T_* = 2.725/a_*$ K. The most stringent constraints at the early universe are derived from primordial nucleosynthesis. Nonsingular models need the critical temperature $T_* > 10^{10}$ K, since light elements He, D, and Li are produced around $T \simeq 10^8 - 10^9$ K in accordance with the observed abundance as described in SBBN. It is easily found from Table 3.2 that when $a_* < 10^{-10}$, this condition can be attained.

Table 3.2 Cosmological values at the moment of the *bounce*

γ	Ω_0	m	a_*	λ_*	T_*(K)
1	0.51	1.5	1.70856E–04	0.510	1.595E+04
1	0.61	1.8	4.75822E–10	0.407	5.727E+09
4/3	0.55	1.9	1.36487E–09	0.608	1.997E*09
4/3	0.54	1.9	3.92015E–10	0.597	6.951E*09
4/3	0.53	1.9	8.89150E–11	0.586	3.065E+10
4/3	0.50	1.9	5.99525E–14	0.553	4.545E+13

When $m \neq 3\gamma$, the energy density is found to be [16]

$$\rho = \rho_0 a^{-3\gamma} \left[1 - \kappa \frac{m(1 - a^{3\gamma - m})}{3\gamma - m} \right], \tag{3.57}$$

where $\kappa = \lambda_0/\Omega_0$. If either $\lambda_0 = 0$ or $m = 0$, then (3.57) reduces to (2.17), and ρ satisfies the thermodynamic condition of the adiabatic expansion of the universe.

We examine thermal evolution during the early stage of the universe. Then we can take $\gamma = 4/3$. Since $\rho \sim T^4$, temperature is written from (3.57) as

$$\frac{T}{T_0} = \frac{1}{a} \left[1 - \kappa \frac{m(1 - a^{4-m})}{4 - m} \right]^{1/4}, \tag{3.58}$$

We guess $0.1 < \kappa < 10$ from observations. We can set $a^{4-m} \ll 1$ at the early universe because $a \simeq 10^{-10}$. Then a necessary condition for $T > 0$ is

$$m < \frac{4}{\kappa + 1}. \tag{3.59}$$

The alternative condition that the right hand side of (3.54) must be negative at the present epoch gives rise to

$$m\kappa < 3. \tag{3.60}$$

Figure 3.10 shows the evolution of temperature in the early universe for several sets of κ and m. It can be seen that temperature becomes lower in a variable Λ model than that of the standard model ($m = 0$).

Fig. 3.10 Temperature in units of its present value T_0 against the scale factor for several sets of κ and m

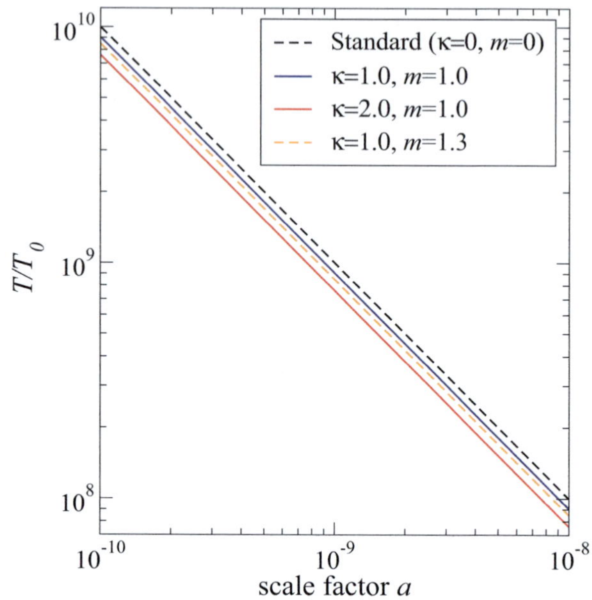

References

1. Wagoner, R.V., Fowler, W.A., Hoyle, F.: On the synthesis of elements at very high temperatures. Astrophys. J. **148**, 3–49 (1967). https://doi.org/10.1086/149126
2. Weinberg, S.: Gravitation and Cosmology: Principles and Applications of the General Theory of Relativity, p. 545. Wiley, New York (1972)
3. Aver, E., Olive, K.A., Skillman, E.D.: The effects of He I λ10830 on helium abundance determinations. J. Cosmol. Astropart. Phys. **07**, 011 (2015). https://doi.org/10.1088/1475-7516/2015/07/011
4. Cooke, R.J., et al.: Precision measures of the primordial abundance of deuterium. Astrophys. J. **781**, 31 (2014). https://doi.org/10.1088/0004-647X/781/1/31
5. Press, W.H., Teukolsky, S.A., Vetterling, W.T., Flannery, B.P.: Numerical Recipes. The Art of Scientific Computing, 3rd ed., p. 815. Cambridge University Press, Cambridge (2007)
6. Maartens, R.: Brane-world gravity. Living Rev. Relativ. **7**, 7 (2004). https://doi.org/10.1007/b99562
7. Randall, L., Sundrum, R.: An alternative to compactification. Phys. Rev. Lett. **83**, 469 (1999). https://doi.org/10.1103/PhysRevLett.83.469
8. Randall, L., Sundrum, R.: A large mass hierarchy from a small extra dimension. Phys. Rev. Lett. **83**, 3370 (1999). https://doi.org/10.1103/PhysRevLett.83.3370
9. Binetruy, P., Deffayet, C., Ellwanger, U., Langlois, D.: Brane cosmological evolution in a bulk with cosmological constant. Phys. Lett. B **477** 285 (2000). https://doi.org/10.1016/S0370-2693(00)00204-5
10. Binetruy, P., Deffayet, C., Langlois, D.: Nonconventional cosmology from a brane universe. Nucl. Phys. B **565**, 269 (2000). https://doi.org/10.1016/S0550-3213(99)00696-3
11. Flanagan, E.E., Tye, S.-H.H., Wasserman, I.: Cosmological expansion in the Randall-Sundrum brane world scenario. Phys. Rev. D **62**, 044039 (2000). https://doi.org/10.1103/PhysRevD.62.044039

12. Ichiki, K., Yahiro, M., Kajino, T., Orito, M., Mathews, G.J.: Observational constraints on dark radiation in brane cosmology. Phys. Rev. D **66**, 043521 (2002). https://doi.org/10.1103/PhysRevD.66.043521
13. Bratt, J.D., Gault, A.C., Scherrer, R.J., Walker, T.P.: Big bang nucleosynthesis constraints on brane cosmologies. Phys. Lett. B **546**, 19 (2002). https://doi.org/10.1016/S0370-2693(02)02637-0
14. Sasankan, N., Gangopadhyay, M.R., Mathews, G.J., Kusakabe, M.: New observational limits on dark radiation in brane world cosmology. Phys. Rev. D **95** 083516 (2017). https://doi.org/10.1103/PhysRevD.95.083516
15. Alcaniz, J.S., Lima, J.A.S.: Interpreting cosmological vacuum decay. Phys. Rev. D **72**, 063516 (2005). https://doi.org/10.1103/PhysRevD.72.063516
16. Overduin, J.M., Cooperstock, F.I.: Evolution of the scale factor with a variable cosmological term. Phys. Rev. D **58**, 043506 (1998). https://doi.org/10.1103/PhysRevD.58.043506

Chapter 4
Modified Brans-Dicke Theory

Abstract Concerning the decrease in the cosmological term Λ from a large magnitude in a very early universe to a small value at the present epoch, we explore the Brans-Dicke theory modified with a variable Λ term which is a function of the scalar field. The fundamental equations are derived for the gravitational and scalar fields from the variational principle of the action. In the framework of the Robertson-Walker metric, we obtain the expression for the expansion rate of the universe like the Friedmann equation. Then we turn to summarize the observational constraints on the intrinsic parameters contained in this theory. We confirm that the evolution of the universe deviates significantly from the standard model in the early stage. Calculations of Big Bang nucleosynthesis (BBN) are performed with the use of the nuclear reaction network. Comparing the resultant amounts of the light elements, ^{4}He, D, and ^{7}Li, with the observed primordial abundances and using also the magnitude-redshift relation of Type Ia supernovae (SNe Ia), we derive reasonable ranges of the parameters.

Keywords Brans-Dicke theory · Flat universe · Scaler field · Variable cosmological term

4.1 Dynamics of BDΛ Model

Based on the original Brans-Dicke (BD) theory [1], we construct a BD model incorporated with a variable Λ term which is a function of the scalar field ϕ (hereafter BDΛ). The action is expressed as [2]

$$S = \int d^4x\sqrt{-g}\left[(R - 2\Lambda)\phi - \frac{\omega}{\phi}\phi_{,\nu}\phi^{,\nu} + 16\pi\mathscr{L}_m\right], \tag{4.1}$$

where $\phi_{,\nu} = \partial\phi/\partial x^\nu$, R is the curvature scalar, $\mathscr{L}_m$ is the Lagrangian density of matter, and ω is the dimensionless constant of BD gravity.

M. Hashimoto et al., *Big-Bang Nucleosynthesis*, SpringerBriefs in Physics,
https://doi.org/10.1007/978-981-13-2935-7_4

The variational principle with respect to $g_{\mu\nu}$ yields the field equation

$$R_{\mu\nu} - \frac{1}{2}g_{\mu\nu}R + g_{\mu\nu}\Lambda = \frac{8\pi}{\phi}T_{\mu\nu} + \frac{\omega}{\phi^2}\left(\phi_{,\mu}\phi_{,\nu} - \frac{1}{2}g_{\mu\nu}\phi_{,\sigma}\phi^{,\sigma}\right) + \frac{1}{\phi}\left(\phi_{,\mu;\nu} - g_{\mu\nu}\Box\phi\right), \tag{4.2}$$

where $\Box$ is the d'Alembertian, the symbol semicolon indicates the covariant differentiation and

$$T^{\mu\nu} = \frac{2}{\sqrt{-g}}\frac{\partial}{\partial g_{\mu\nu}}\left[\sqrt{-g}\mathscr{L}_m\right] \tag{4.3}$$

is the energy-momentum tensor of matter.

Contraction of (4.2) results in

$$-R + 4\Lambda = \frac{8\pi}{\phi}T^{\nu}_{\nu} - \frac{\omega}{\phi^2}\phi_{,\nu}\phi^{,\nu} - \frac{3}{\phi}\Box\phi. \tag{4.4}$$

On the other hand, the field equation for ϕ is obtained from the variational principle with respect to ϕ in (4.1) as

$$R - 2\Lambda - 2\phi\frac{\partial\Lambda}{\partial\phi} = \frac{\omega}{\phi^2}\phi_{,\nu}\phi^{,\nu} - \frac{2\omega}{\phi}\Box\phi. \tag{4.5}$$

Eliminating R from (4.4) and (4.5), we get

$$\Lambda - \phi\frac{\partial\Lambda}{\partial\phi} = \frac{4\pi}{\phi}T^{\nu}_{\nu} - \frac{2\omega + 3}{2\phi}\Box\phi. \tag{4.6}$$

We adopt the simplest case of coupling between the scalar and matter fields

$$\Box\phi = \frac{8\pi\mu}{2\omega + 3}T^{\nu}_{\nu}, \tag{4.7}$$

where μ is a constant. Then (4.6) is integrated to be

$$\Lambda = \frac{2\pi(1 - \mu)}{\phi}T^{\nu}_{\nu}. \tag{4.8}$$

It is noted that the case $\mu = 1$ reduces to the original BD theory with $\Lambda = 0$.

Here we consider the homogeneous and isotropic universe, as depicted by the Cosmological Principle in Chap. 1. It follows that the scalar field ϕ turns out to be a function of time only. Using the Robertson-Walker metric (2.14) and the energy-momentum tensor of perfect fluid (2.9), after some manipulations, the (0, 0) component of (4.2) is written as

$$\left(\frac{\dot{a}}{a}\right)^2 + \frac{k}{a^2} - \frac{\Lambda}{3} - \frac{\omega}{6}\left(\frac{\dot{\phi}}{\phi}\right)^2 + \frac{\dot{a}}{a}\frac{\dot{\phi}}{\phi} = \frac{8\pi}{3}\frac{\rho}{\phi}. \tag{4.9}$$

Consequently, the expansion rate of the BDΛ model is given by

$$H = \frac{\dot{a}}{a} = \left[\frac{1}{4}\left(\frac{\dot{\phi}}{\phi}\right)^2 - \frac{k}{a^2} + \frac{\Lambda}{3} + \frac{\omega}{6}\left(\frac{\dot{\phi}}{\phi}\right)^2 + \frac{8\pi}{3}\frac{\rho}{\phi}\right]^{1/2} - \frac{\dot{\phi}}{2\phi}, \tag{4.10}$$

where we adopt an expanding solution.

For a perfect fluid we obtain from (4.7)

$$\frac{d}{dt}\left(\dot{\phi}a^3\right) = \frac{8\pi\mu}{2\omega+3}(\rho - 3p)\,a^3. \tag{4.11}$$

Since $\rho = \rho_m + \rho_r$ and $p = p_r = \rho_r/3$, (4.11) is integrated to give

$$\dot{\phi} = \frac{1}{a^3}\left[\frac{8\pi\mu}{2\omega+3}\rho_{m0}t + B\right], \tag{4.12}$$

where B is an integral constant, ρ_{m0} is the matter density at the present epoch, and the scale factor a is normalized to its present value ($a_0 = 1$).

From (4.8) we have

$$\Lambda = \frac{2\pi(\mu - 1)}{\phi}\rho_m. \tag{4.13}$$

The gravitational "constant" G in the BDΛ model can be expressed as

$$G = \frac{1}{2}\left(3 - \frac{2\omega+1}{2\omega+3}\mu\right)\frac{1}{\phi}. \tag{4.14}$$

When $\omega \to \infty$ and $\mu = 1$, we obtain $G = 1/\phi$. Moreover, if ϕ is constant, then (4.9) in our model reduces to the Friedmann equation (2.21) without the Λ term.

The curvature constant k is given from (4.9) evaluated at the present epoch as

$$k = \frac{2\pi(5-\mu)}{3\phi_0}\rho_{m0} - H_0{}^2 + \frac{\omega}{6}\left(\frac{\dot{\phi}}{\phi}\right)_0 - H_0\left(\frac{\dot{\phi}}{\phi}\right)_0. \tag{4.15}$$

4.2 Parameters in the BDΛ Model

As described in the previous section, our BDΛ model contains three parameters, ω, μ, and B. It can be seen from (4.14) that the strength of the ϕ field is related to the gravitational "constant."

The coupling constant was proposed to be $\omega = 6$ in the original BD theory [1]. BBN has been studied for $\omega = 6$ in the BDΛ model [3]. A detailed analysis [4] gives $\omega > 50$. Using the Viking space probe, time delay of radar echoes results in $\omega > 500$ [5]. Calculations of BBN are revisited with $\omega = 500$ [6]. From the Cassini spacecraft, the result $\omega > 40{,}000$ is reported [7]. It is, therefore, worthwhile to investigate the BDΛ model with the use of a new value of ω. Here we adopt $\omega = 10{,}000$ in our studies [8].

The time variation of the gravitational "constant" G has been explored using various techniques. The results are summarized in Table 4.1 for the limit to $(\dot{G}/G)_0$ at the present epoch obtained from, e.g., palaeomagnetic studies [9], lunar laser ranging [11], neutron star masses [13], and helioseismology [15]. Consequently, we set the constraint $(\dot{G}/G)_0 < 10^{-13}\,\mathrm{yr}^{-1}$.

As can be seen from (4.13), if the intrinsic parameter $\mu < 1$, we have $\Lambda < 0$. Also, from (4.14), we obtain $\phi G < 0$ if $\mu > 3$ and $\omega \gg 1$. Therefore, it is suitable to consider the range $-2 \le \mu \le 2$.

We can see in (4.12) that the behavior of the early universe is essentially determined from B as

$$\lim_{t\to 0} \dot{\phi}(t) = \lim_{t\to 0} \frac{B}{a^3(t)}.$$

A model with negative B implies negative $\dot{\phi}$, or equivalently positive $\dot{G}$ in the early stage of the universe.

On the other hand, using $t_0 \simeq 10^{17}$ s and $\rho_{m0} \simeq 10^{-31}\,\mathrm{g\,cm^{-3}}$, we can find that the BDΛ model does not deviate appreciably from the standard model near the present epoch as long as $|B^*| \le 10$, where B^* is the normalized value of B defined as $B^* = B/(10^{-24}\,\mathrm{g\,s\,cm^{-3}})$. Then we obtain

Table 4.1 Time variation of the gravitational "constant" $(\dot{G}/G)_0$

Method	$(\dot{G}/G)_0$ $(10^{-12}\mathrm{yr}^{-1})$	Reference
Palaeomagnetic studies	<8	[9]
Big bang nucleosynthesis	<0.17	[10]
Radar ranging of Mars	0.8	[5]
Lunar laser ranging	0.1 ± 10.4	[11]
White dwarf	$-30(^{+10}_{-30})$	[12]
Neutron star mass	-0.6 ± 2.0	[13]
Globular cluster age	$-35 \sim 7$	[14]
Helioseismology	<1.6	[15]

Table 4.2 Parameters in the BDΛ model

Parameter		Value
Coupling constant	ω	10,000
Intrinsic parameter	μ	$-2 \sim 2$
Integral constant	B^*	$-10 \sim 10$

$$(\dot{\phi})_0 = \frac{8\pi\mu}{2\omega+3}\rho_{m0}t_0.$$

Using (4.14), we have

$$\left(\frac{\dot{G}}{G}\right)_0 = -\left(\frac{\dot{\phi}}{\phi}\right)_0 = -\frac{16\pi\rho_{m_0}G_0t_0}{3(2\omega+3)-(2\omega+1)\mu}. \tag{4.16}$$

Table 4.2 gives the values of parameters in the BDΛ model we adopt in our numerical calculations.

4.3 Characteristics of a Flat Universe

We examine characteristics of the BDΛ model in a flat universe. The Hubble parameter at the present epoch is expressed from (4.9) with $k = 0$ as

$$H_0^2 = \frac{8\pi\rho_{m0}}{3\phi_0} + \frac{\Lambda_0}{3} + \frac{\omega}{6}\left(\frac{\dot{\phi}}{\phi}\right)_0^2 - H_0\left(\frac{\dot{\phi}}{\phi}\right)_0. \tag{4.17}$$

Let us introduce density parameters

$$\Omega_{m0} = \frac{\rho_{m0}}{\rho_{cr}^{BD\Lambda}}, \qquad \Omega_{\Lambda 0} = \frac{\Lambda_0}{3H_0^2},$$

$$\Omega_{\phi 0} = \frac{\omega}{6H_0^2}\left(\frac{\dot{\phi}}{\phi}\right)_0^2 - \frac{1}{H_0}\left(\frac{\dot{\phi}}{\phi}\right)_0,$$

where $\rho_{cr}^{BD\Lambda} = 3\phi_0 H_0^2/8\pi$ is the critical density of the BDΛ model. Then we have

$$\Omega_{m0} + \Omega_{\Lambda 0} + \Omega_{\phi 0} = 1, \tag{4.18}$$

Since $|\dot{\phi}/\phi|_0 = |\dot{G}/G|_0 < 10^{-12}\ \mathrm{yr}^{-1}$ (see Table 4.1), we get $\Omega_{\phi 0} < 10^{-2}$. Using (4.13), we obtain

$$\Omega_{m0} \simeq \frac{4}{\mu+3}, \qquad \Omega_{\Lambda 0} \simeq \frac{\mu-1}{\mu+3}.$$

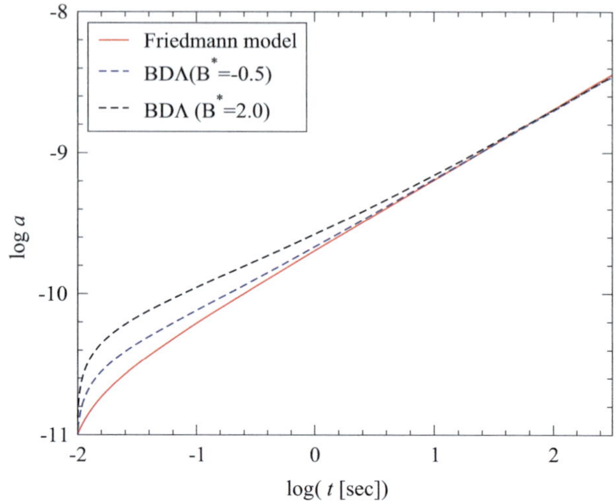

Fig. 4.1 Evolution of the scale factor in the BDΛ model for $B^* = -0.5$ and 2.0 with $\mu = 0.6$ and $\omega = 10^4$. The solid line denotes the standard model

Therefore, the matter density at the present epoch is given by

$$\rho_{m0} = \frac{3}{2\pi(\mu + 3)}\phi_0 H_0^2. \tag{4.19}$$

We can solve a set of (4.10), (4.12), and (4.13) numerically with specified parameters ω, μ, and B^*. The others are fixed to be $H_0 = 71\,\mathrm{km\,s^{-1}\,Mpc^{-1}}$ and $G_0 = 6.6726 \times 10^{-8}\,\mathrm{dyne\,cm^2\,g^{-2}}$. The matter density ρ_{m0} at the present epoch is evaluated from (4.19).

Figure 4.1 shows the evolution of the scale factor in the BDΛ model for $B^* = -0.5$ and 2.0 with fixed $\mu = 0.6$ and $\omega = 10^4$. The deviations from the standard model, which are indicated by the solid line, become appreciable at $t < 10\,\mathrm{s}$ in the early stage of the universe.

As already stated, $\dot{\phi}$ is essentially determined from B^* in the early universe. The evolution of the scalar field is shown in Fig. 4.2 for several values of ω and B^* with $\mu = 0.6$ fixed. As can be seen from (4.12), if B^* is positive, ϕ is an increasing function of time. It follows that G is a decreasing function. In contrast, if B^* is negative, ϕ decreases until $t \simeq 10\,\mathrm{s}$. Thereafter, the first term in (4.12) becomes dominant over $|B|$, and $\dot{\phi}$ changes to positive. Consequently, there is no considerable difference in the BDΛ model from B^* at $t > 10^2\,\mathrm{s}$.

Figure 4.3 shows the evolution of the scalar field for several values of ω and μ with $B^* = 0.5$ fixed. Since B^* is positive, ϕ increases monotonically with time. For large ω, (4.14) reduces to

$$\phi \simeq \frac{3-\mu}{2}\frac{1}{G}.$$

Then we have large ϕ for small μ.

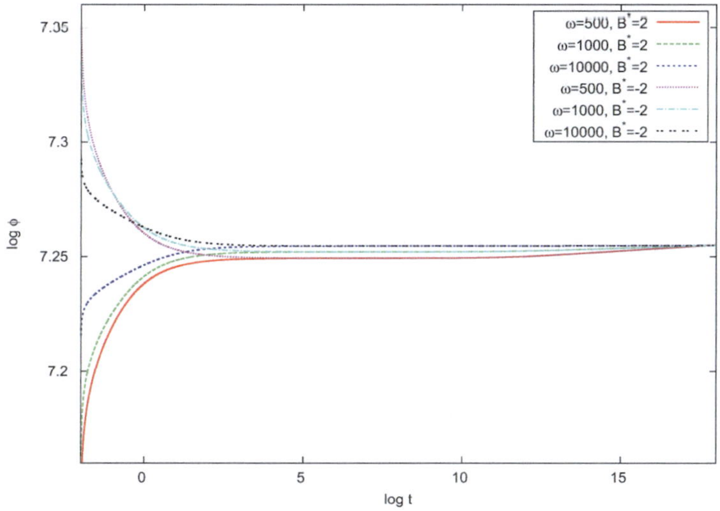

Fig. 4.2 Evolution of the scalar field in the BDΛ model for several values of ω and B^* with $\mu = 0.6$

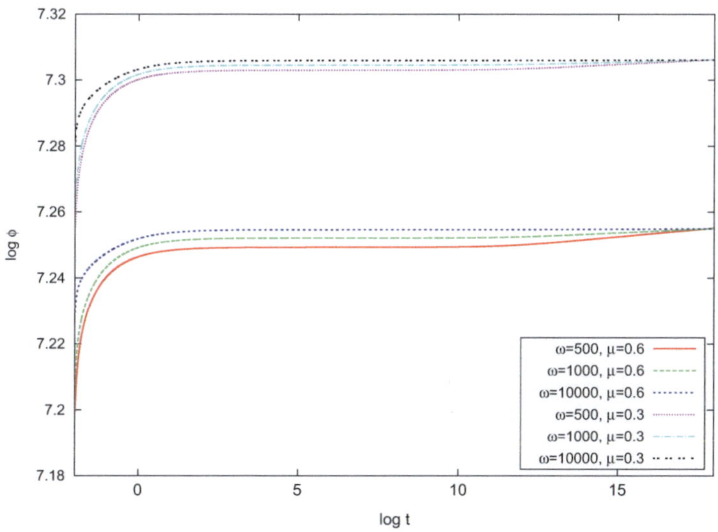

Fig. 4.3 Evolution of the scalar field in the BDΛ model for several values of ω and μ with $B^* = 0.5$

As can be seen from Figs. 4.1, 4.2, and 4.3, the BDΛ model deviates from the standard model. The deviation, which depends on the specified parameters, is considerably large at $t < 100$ s. Therefore, these parameters can be constrained from observational points of view such as the abundance of light elements in BBN and the $m - z$ relation of SNIa.

4.4 Constraints on the BDΛ Model

4.4.1 Constraints from BBN

Synthesis of ^{4}He is the most important consequence of BBN. Its abundance should be in principle used as the critical test of nonstandard models, because it reflects physical circumstances at around $T \simeq 10^9$ K in the early universe. The abundance of light elements in the BDΛ model has already been investigated [3, 6, 16], though the parameter ω was taken to be small. We perform BBN calculations for $\omega = 10^4$ using the nuclear reaction network as described in Sect. 2.5 of SBBN.

Figure 4.4 shows the produced abundance of ^{4}He, D, and ^{7}Li against the baryon-to-photon ratio η. We take $B^* = -10, -2.5, 0$, and 10 with $\mu = 0.5$ fixed. As can be seen from Fig. 4.1, when $|B^*|$ is large, the expansion rate of the universe increases, which leads to more neutrons at the onset of nucleosynthesis. Consequently, more ^{4}He is produced than in SBBN.

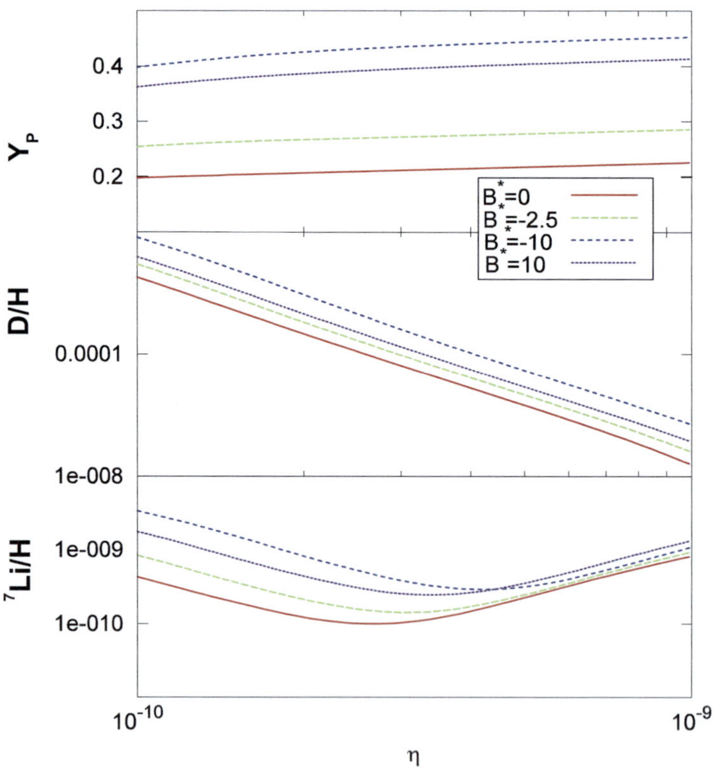

Fig. 4.4 Abundance of light elements produced in the BDΛ model for various values of B^* with $\mu = 0.5$ and $\omega = 10^4$

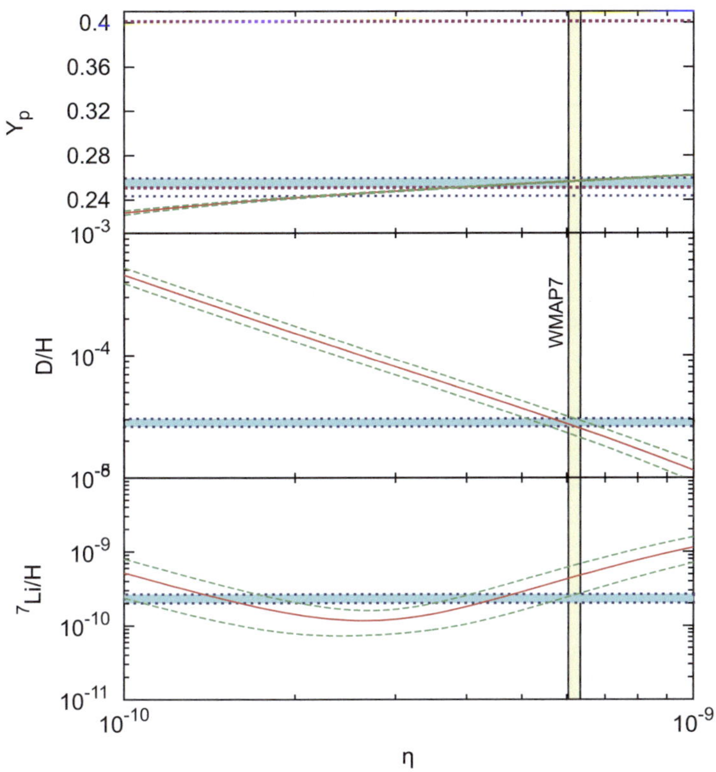

Fig. 4.5 Abundance of light elements produced in the BDΛ model for $B^* = 2$, $\mu = 0.6$, and $\omega = 10^4$. The dashed lines indicate the 2σ uncertainties in the nuclear reaction rates. The horizontal dotted lines denote the observed abundances of Y_p [17–19], D/H [20], and ^{7}Li/H [21]. The vertical lines indicate the range of η determined from WMAP [22]

We compare the calculated yields of BBN with the observed primordial abundances. Figure 4.5 shows the produced abundance for $B^* = 2$ and $\mu = 0.6$. The dashed lines denote the 2σ uncertainties in the nuclear reaction rates. The observed abundances are indicated by the horizontal dotted lines. The data are taken from Refs. [17–20] and [21] for Y_p, D/H, and ^{7}Li/H, respectively. The vertical lines indicate the value $\eta = (6.19 \pm 0.15) \times 10^{-10}$ determined from WMAP [22].

It is found that the range of η derived from both Y_p and D/H is tightly consistent with the value of WMAP, though the lower limit of ^{7}Li/H is barely safe. Finally, we obtain the reasonable parameter ranges

$$0.0 \leq \mu \leq 0.6, \qquad -2 \leq B^* \leq 2. \tag{4.20}$$

4.4.2 Constraints from SNe Ia

Now we proceed to consider the magnitude-redshift ($m - z$) relation in our BDΛ model with $k = 0$ as described in Sect. 2.6 for the standard model. We are concerned with a matter dominant era of $z < 5$, where the parameter B^* is not effective to change the evolution of the universe.

Once the parameters ω, μ, and B^* are specified, the expansion rate (4.9) and hence the luminosity distance (2.58) are numerically evaluated. The resulting m-z relation is shown in Fig. 4.6 together with the observed data for SNe Ia taken from the Union2 compilation [23].

We make the same χ^2 analysis for the distance moduli of SNe Ia as in Sect. 2.6. We get $\chi^2 = 2293$ for the samples of $N = 557$. Hence the reduced χ-square becomes $\chi_r^2 = \chi^2/N = 4.12$. When $\dot{\phi} \simeq 0$ in the matter dominated stage, we have $\Lambda \sim \rho_m$ from (4.13). It follows that the BDΛ model reduces to the standard model without Λ. This is contradictory to the present accelerating universe, which should contain a sufficient amount of dark energy to accelerate the universe. To simulate the present acceleration, we are forced to add another constant cosmological term Λ_{c0}.

The expansion rate in this model is given by

$$H = \left[\frac{1}{4}\left(\frac{\dot{\phi}}{\phi}\right)^2 + \frac{\Lambda}{3} + \frac{\Lambda_{c0}}{3} + \frac{\omega}{6}\left(\frac{\dot{\phi}}{\phi}\right)^2 + \frac{8\pi}{3}\frac{\rho}{\phi}\right]^{1/2} - \frac{\dot{\phi}}{2\phi}. \tag{4.21}$$

When we set $\Omega_{\Lambda c0} = \Lambda_{c0}/(3H_0^2) = 0.7$ and $\mu = 0.5$, the resultant m-z relation is shown in Fig. 4.6. We obtain $\chi^2 = 546.9$ and hence $\chi_r^2 = 0.98$. It is noted

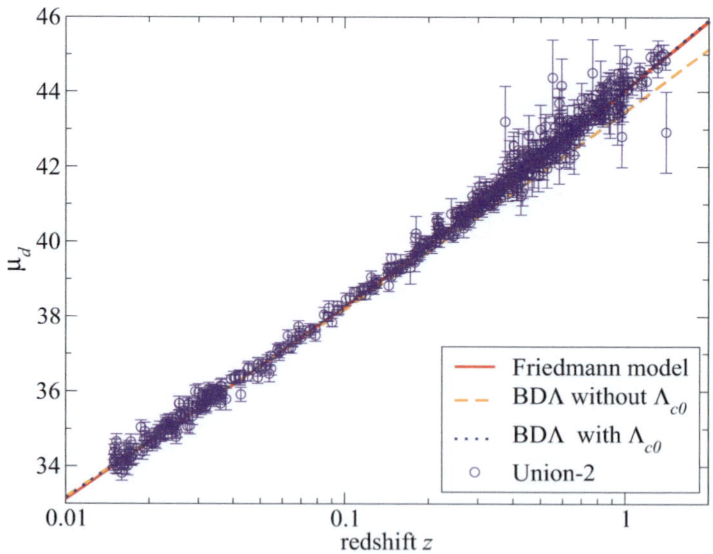

Fig. 4.6 Magnitude-redshift relation of a flat universe in the standard and BDΛ models

that $\Omega_{\Lambda 0} = -0.043$ and $\Omega_{m0} = 0.34$ in this model. Consequently, the total Ω_Λ amounts to 0.66. We conclude that the BDΛ model with Λ_{c0} has nearly the same characteristics as the standard model with $\Omega_m = 0.3$ and $\Omega_\Lambda = 0.7$. Although the cosmological term is not important in the early stage, it plays a very important role near the present epoch.

References

1. Brans, C., Dicke, R.H.: Mach's principle and a relativistic theory of gravitation. Phys. Rev. **124**, 925–935 (1961). https://doi.org/10.1103/PhysRev.124.925
2. Endo, M., Fukui, T.: The cosmological term and a modified Brans-Dicke cosmology. Gen. Rel. Grav. **8**, 833–839 (1977). https://doi.org/10.1007/BF00759587
3. Arai, K., Hashimoto, M., Fukui, T.: Primordial nucleosynthesis in the Brans-Dicke theory with a variable cosmological term. Astron. Astrophys. **179**, 17–22 (1987)
4. Liddle, A.R., Lyth, H.: Cosmological inflation and large-scale structure. Cambridge University Press, Cambridge (2000)
5. Shapiro, I.I.: Solar system tests of general relativity: recent results and present plans. In: Ashby, N., Bartlett, D.F., Wyss, W. (eds.) General Relativity and Gravitation, p. 313. Cambridge University Press, Cambridge (1990)
6. Etoh, T., Hashimoto, M., Arai, K., Fujimoto, S.: Age of the universe constrained from the primordial nucleosynthesis in the Brans-Dicke theory with a varying cosmological term. Astron. Astrophys. **325**, 893–897 (1997)
7. Bertotti, B., Iess, L., Tortora, P.: A test of general relativity using radio links with the Cassini spacecraft. Nature **425**, 374–376 (2003). https://doi.org/10.1038/nature01997
8. Thushari, E.P.B.A., Nakamura, R., Hashimoto, M., Arai, K.: Brans-Dicke model constrained from the big bang nucleosynthesis and magnitude redshift relations of supernovae. Astron. Astrophys. (Research Note) **521**, A52–A55 (2010)
9. McElhinny, M.W., Rayloran, S.R., Stevenson, D.J.: Limits to the expansion of Earth, Moon, Mars and mercury and to changes in the gravitational constant. Nature **271**, 316–321 (1978). https://doi.org/10.1038/271316a0
10. Rothman, T., Matzner, R.: Scale-covariant gravitation and primordial nucleosynthesis. Astrophys. J. **257**, 450–455 (1982). https://doi.org/10.1086/160003
11. Müller, J., Schneider, M., Soffel, M., Ruder, H.: Testing Einstein's theory of gravity by analyzing lunar laser ranging data. Astrophys. J. **382**, L101–L103 (1991). https://doi.org/10.1086/186222
12. Garcia-Berro, E., Hernanz, M., Isern, J., Mochkovitch, R.: The rate of change of the gravitational constant and the cooling of white dwarfs. Mon. Not. R. Astron. Soc. **277**, 801–810 (1995). https://doi.org/10.1093/mnras/277.3.801
13. Thorsett, S.E.: The gravitational constant, the Chandrasekhar limit, and the neutron star masses. Phys. Rev. Lett. **77**, 1432–1435 (1996)
14. Degl'Innocenti, S., Fiorentini, G., Raffelt, G.G., Ricci, B., Weiss, A.: Time-variation of Newton's constant and the age of globular clusters. Astron. Astrophys. **312**, 345–352 (1996)
15. Guenther, D.B., Krause, L.M., Demarque, P.: Testing the constancy of the gravitational constant using helioseismology. Astrophys. J. **498**, 871–876 (1998). https://doi.org/10.1086/305567
16. Nakamura, R., Hashimoto, M., Gamow, S., Arai, K.: Big-bang nucleosynthesis in a Brans-Dicke cosmology with a varying Λ term related to WMAP. Astron. Astrophys. **448**, 23–27 (2006). https://doi.org/10.1051/0004-6361:20042618
17. Izotov, Y.I., Thuan, T.X.: Systematic effects and a new determination of the primordial abundance of 4He and dY/dZ from observations of blue compact galaxies. Astrophys. J. **602**, 200 (2004). https://doi.org/10.1086/380830

18. Izotov, Y.I., Thuan, T.X., Stasinska, G.: The primordial abundance of 4He: a self-consistent empirical analysis of systematic effects in a large sample of low-metallicity H II regions. Astrophys. J. **662**, 15 (2007). https://doi.org/10.1086/513601
19. Peimbert, M., Luridiana, V., Peimbert, A.: Revised primordial helium abundance based on new atomic data. Astrophys. J. **666**, 636 (2007). https://doi.org/10.1086/520571
20. Pettini, M., et al.: Deuterium abundance in the most metal-poor damped Lyman alpha system: converging on $\Omega_{b,0}h^2$. Mon. Not. R. Astron. Soc. **391**, 1499–1510 (2008). https://doi.org/10.1111/j.1365-2966.2008.13921.x
21. Melendez, J., Ramirez, I.: Reappraising the spite lithium plateau: extremely thin and marginally consistent with WMAP data. Astrophys. J. **615**, L33 (2004). https://doi.org/10.1086/425962
22. Komatsu, E., et al.: Seven-year Wilkinson microwave anisotropy probe (WMAP) observations: cosmological interpretation. Astrophys. J. Suppl. **192**, 18 (2011). https://doi.org/10.1088/0067-0049/192/2/18
23. Amanullah, R., et al. (The supernova cosmology project): Spectra and hubble space telescope light curves of six type Ia supernovae at $0.511 < z < 1.12$ and the Union2 compilation. Astrophys. J. **716**, 712 (2010). https://doi.org/10.1088/0004-637X/716/1/712

Chapter 5
Appendix A: Einstein Tensor

Abstract We provide some useful formulas explicitly to obtain the Einstein tensor as a convenience to readers. We also establish our notation in tensor calculus.

Keywords Einstein tensor · Notation in tensor calculus

The infinitesimal distance ds of two points with difference dx^μ is given by

$$ds^2 = g_{\mu\nu} dx^\mu dx^\nu, \tag{5.1}$$

where $g_{\mu\nu}$ is the metric tensor. The Christoffel symbol is written as

$$\Gamma^\lambda_{\mu\nu} = \frac{1}{2} g^{\lambda\sigma} \left(\frac{\partial g_{\sigma\mu}}{\partial x^\nu} + \frac{\partial g_{\sigma\nu}}{\partial x^\mu} - \frac{\partial g_{\mu\nu}}{\partial x^\sigma} \right). \tag{5.2}$$

The variational principle under the condition that the endpoints A and B are fixed

$$\delta \int_{\mathrm{A}}^{\mathrm{B}} ds = 0$$

yields the geodesic equation

$$\frac{d^2 x^\lambda}{d\tau^2} + \Gamma^\lambda_{\mu\nu} \frac{dx^\mu}{d\tau} \frac{dx^\nu}{d\tau} = 0, \tag{5.3}$$

where τ is the proper time defined as $ds^2 = -d\tau^2$.

The covariant derivative of a contravariant vector is

$$A^\mu{}_{;\nu} = \frac{\partial A^\mu}{\partial x^\nu} + \Gamma^\mu_{\nu\lambda} A^\lambda, \tag{5.4}$$

M. Hashimoto et al., *Big-Bang Nucleosynthesis*, SpringerBriefs in Physics,
https://doi.org/10.1007/978-981-13-2935-7_5

and similarly the covariant derivative of a covariant vector is

$$A_{\mu;\nu} = \frac{\partial A_\mu}{\partial x^\nu} - \Gamma^\lambda_{\mu\nu} A_\lambda. \tag{5.5}$$

More generally, the covariant derivative of a tensor ${T^{\mu\nu}}_\lambda$ is

$${T^{\mu\nu}}_{\lambda;\rho} = \frac{\partial {T^{\mu\nu}}_\lambda}{\partial x^\rho} + \Gamma^\mu_{\rho\sigma} {T^{\sigma\nu}}_\lambda + \Gamma^\nu_{\rho\sigma} {T^{\mu\sigma}}_\lambda - \Gamma^\sigma_{\lambda\rho} {T^{\mu\nu}}_\sigma. \tag{5.6}$$

The Riemann tensor is given by

$${R^\lambda}_{\rho\mu\nu} = \frac{\partial \Gamma^\lambda_{\rho\nu}}{\partial x^\mu} - \frac{\partial \Gamma^\lambda_{\rho\mu}}{\partial x^\nu} + \Gamma^\lambda_{\sigma\mu}\Gamma^\sigma_{\rho\nu} - \Gamma^\lambda_{\sigma\nu}\Gamma^\sigma_{\rho\mu}. \tag{5.7}$$

Its contraction yields the Ricci tensor as

$$\begin{aligned} R_{\mu\nu} &= {R^\lambda}_{\mu\lambda\nu} \\ &= \frac{\partial \Gamma^\lambda_{\mu\nu}}{\partial x^\lambda} - \frac{\partial \Gamma^\lambda_{\mu\lambda}}{\partial x^\nu} + \Gamma^\lambda_{\mu\nu}\Gamma^\rho_{\lambda\rho} - \Gamma^\lambda_{\mu\rho}\Gamma^\rho_{\nu\lambda}. \end{aligned} \tag{5.8}$$

Moreover contracted is the curvature scalar

$$R = g^{\mu\nu} R_{\mu\nu} = {R^\mu}_\mu. \tag{5.9}$$

The Einstein tensor is given by

$$G_{\mu\nu} = R_{\mu\nu} - \frac{1}{2} R\, g_{\mu\nu}. \tag{5.10}$$

Chapter 6
Appendix B: Thermal History in an Early Universe

Abstract We describe the thermal history during the stage of annihilation of electron-positron pairs. Energies of the pairs are converted into photons. We obtain the expression for the photon temperature T which deviates from the usual law $T \sim a^{-1}$ in the expanding universe.

Keywords Thermal history · Photon temperature

Let us consider the thermal history during the stage when electron-positron pairs annihilate and hence convert their energy into photons. In the early universe, the Friedmann equation (2.21) can be written as

$$\frac{\dot{a}}{a} = \left(\frac{8\pi G}{3}\rho_r\right)^{1/2}.$$

The constituents of radiation at $T \simeq 10^{10}$ K are photons, neutrinos, and electron-positron pairs which are mildly relativistic. The energy density is given by

$$\rho_r = \rho_\gamma + \rho_\nu + \rho_e.$$

Here

$$\rho_\gamma = a_\mathrm{B} T^4, \quad \rho_\nu = \frac{7}{8} N_\nu a_\mathrm{B} {T_\nu}^4,$$

where a_B is the radiation density constant, $N_\nu = 3$ is the number of neutrino species, and T_ν is the neutrino temperature. Since neutrinos have already been decoupled at this stage, we have $T_\nu \neq T$.

The energy density and pressure of electrons are, neglecting degeneracy, given by

$$\rho_{e^-} = \frac{8\pi}{h^3}\int_0^\infty \frac{Ep^2}{\exp(E/k_\mathrm{B}T)+1}dp, \tag{6.1}$$

M. Hashimoto et al., *Big-Bang Nucleosynthesis*, SpringerBriefs in Physics,
https://doi.org/10.1007/978-981-13-2935-7_6

$$p_{e^-} = \frac{8\pi}{3h^3}\int_0^\infty \frac{p^2}{E}\frac{p^2}{\exp(E/k_{\rm B}T)+1}dp \tag{6.2}$$

with

$$E = (p^2 + {m_e}^2)^{1/2},$$

where h is the Planck constant, $k_{\rm B}$ is the Boltzmann constant, E is the energy of an electron with momentum p, and m_e is the electron rest mass. Recall that we are using units with $c = 1$.

If we take the transformation $p = m_e \sinh\theta$, we have $E = m_e \cosh\theta$. Thus (6.1) and (6.2) can be written as

$$\rho_{e^-} = a_e \int_0^\infty \frac{\sinh^2\theta \cosh^2\theta}{\exp(\zeta\cosh\theta)+1}d\theta, \tag{6.3}$$

$$p_{e^-} = \frac{a_e}{3}\int_0^\infty \frac{\sinh^4\theta}{\exp(\zeta\cosh\theta)+1}d\theta \tag{6.4}$$

where

$$\zeta = \frac{m_e}{k_{\rm B}T}, \quad a_e = \frac{8\pi {m_e}^4}{h^3}.$$

The integrals result in

$$\begin{aligned}
F_1 &= \int_0^\infty \sinh^2\theta[1+\exp(\zeta\cosh\theta)]^{-1}d\theta \\
&= \int_0^\infty \sum_{n=1}(-1)^{n+1}\sinh^2\theta\exp[-n\zeta\cosh\theta]d\theta \\
&= \sum_{n=1}(-1)^{n+1}\frac{1}{n\zeta}K_1(n\zeta), \\
F_2 &= \int_0^\infty \sinh^4\theta[1+\exp(\zeta\cosh\theta)]^{-1}d\theta \\
&= \sum_{n=1}(-1)^{n+1}\frac{3}{(n\zeta)^2}K_2(n\zeta),
\end{aligned}$$

where K_n $(n = 1, 2)$ are the modified Bessel functions. Therefore, we obtain

$$\rho_e = \rho_{e^-} + \rho_{e^+} = 2a_e(F_1 + F_2), \tag{6.5}$$

$$p_e = p_{e^-} + p_{e^+} = \frac{2}{3}a_e F_2 \tag{6.6}$$

The energy conservation law (2.15) is now written as

$$\frac{d}{dT}(\rho_\gamma + \rho_\nu + \rho_e) + (\rho_\gamma + \rho_\nu + \rho_e + p_\gamma + p_\nu + p_e)\frac{3}{a}\frac{da}{dT} = 0, \tag{6.7}$$

where the variable is changed from time t to temperature T. The contribution from neutrinos vanishes because $T_\nu \sim a^{-1}$. It follows that

$$\frac{d\ln a}{d\ln T} = -\frac{1}{3(\rho_\gamma + \rho_e + p_\gamma + p_e)}\frac{d(\rho_\gamma + \rho_e)}{d\ln T}. \tag{6.8}$$

We obtain from (6.2)

$$\frac{dp_e}{d\ln T} = -\frac{dp_e}{d\ln\zeta} = (\rho_e + p_e).$$

Thus, we have

$$\frac{d\rho_e}{d\ln T} = T^4\frac{d}{d\ln T}\left(\frac{\rho_e + p_e}{T^4}\right) + 3(\rho_e + p_e).$$

Because

$$\frac{d\rho_\gamma}{d\ln T} = 4\rho_\gamma = 3(\rho_\gamma + p_\gamma),$$

(6.8) reduces to

$$\frac{d\ln a}{d\ln T} = -1 - \frac{T^4}{3(\rho_\gamma + \rho_e + p_\gamma + p_e)}\frac{d}{d\ln T}\left(\frac{\rho_\gamma + \rho_e + p_\gamma + p_e}{T^4}\right),$$

which is integrated to be

$$\ln a + \ln T + \frac{1}{3}\ln[(\rho_\gamma + \rho_e + p_\gamma + p_e)/T^4] = \text{const}.$$

Therefore, we obtain

$$\log a = -\log T - \frac{1}{3}\log\left[\frac{\rho_e + p_e}{T^4} + \frac{4}{3}a_{\mathrm{B}}\right] + C, \tag{6.9}$$

where C is an integration constant.

The entropy density of photons is given by

$$s = \frac{\rho_\gamma + p_\gamma}{T} = \frac{4}{3}a_{\mathrm{B}}T^3. \tag{6.10}$$

Table 6.1 Relation between the scale factor a/a_0, photon temperature T, and energy density of electrons ρ_e

a/a_0	T (K)	ρ_e (erg cm^{-3})
1.94549E–10	1.00000E+10	1.43386E+05
2.46042E–10	7.94328E+09	5.61462E+04
3.11946E–10	6.30957E+09	2.17608E+04
3.97000E–10	5.01187E+09	8.29547E+03
5.08019E–10	3.98107E+09	3.08047E+03
6.54998E–10	3.16228E+09	1.09831E+03
8.52590E–10	2.51189E+09	3.68363E+02
1.12142E–09	1.99526E+09	1.13054E+02
1.48720E–09	1.58489E+09	3.06413E+01
1.97460E–09	1.25893E+09	7.01540E+00
2.59657E–09	1.00000E+09	1.28417E+00
3.35606E–09	7.94328E+08	1.75539E–01
4.27120E–09	6.30957E+08	1.64577E–02
5.39250E–09	5.01187E+08	9.51606E–04
6.79188E–09	3.98107E+08	2.97208E–05
8.55067E–09	3.16228E+08	4.24757E–07

While, at the epoch $k_B T_1 \gg m_e$, neutrinos, photons, and electrons are in thermal equilibrium, the entropy density of photons plus electron-positron pairs is

$$s_1 = \left(1 + \frac{7}{4}\right) \frac{4}{3} a_B {T_1}^3.$$

The entropy in a volume a^3, which is conserved due to adiabatic expansion, is written as

$$S = sa^3 = s_1 {a_1}^3.$$

Because the neutrino temperature varies as $T_\nu \sim a^{-1}$, the photon temperature well after pair annihilation is increased to be

$$T = \left(\frac{11}{4}\right)^{1/3} T_\nu. \tag{6.11}$$

In Table 6.1 we give the relations between the scale factor a/a_0, the photon temperature T, and the energy density of electrons $\rho_e(T)$ during the stage of annihilation of electron-positron pairs.

Chapter 7
Appendix C: Numerical Approach to Calculate Nucleosynthesis

Abstract We describe a numerical method to calculate the evolution of the abundance of chemical elements. A set of differential equations are reduced to the full-implicit backward difference equations for a finite time interval. We can obtain a set of solutions without meeting numerical instability.

Keywords Abundance evolution · Numerical method

We present briefly a numerical method for calculating the evolution of chemical elements, e.g., BBN based on [1].

Let Y_j (mol g^{-1}) be the abundance of the j-th nucleus, and then the time variation of abundance, for example, Y_1, is determined from the following differential equation:

$$\frac{dY_1}{dt} = \sum_{\substack{12\cdots k\\ lm\cdots n}} \lambda^{12\cdots k}_{lm\cdots n} Y_l Y_m \cdots Y_n - \sum_{\substack{lm\cdots n\\ 12\cdots k}} \lambda^{lm\cdots n}_{12\cdots k} Y_1 Y_2 \cdots Y_k. \tag{7.1}$$

Here the first and second terms indicate the production and destruction of Y_1, respectively, and $\lambda^{lm\cdots n}_{12\cdots k}$ is the rate of the reaction $1+2+\cdots+k \longrightarrow l+m+\cdots+n$, where $1, 2, \cdots, n$ stand for nuclear species.

We represent the thermonuclear reaction rate of $k + 1$ bodies by $N_A^k \langle\sigma v\rangle_{12\cdots k}$, where N_A is Avogadro's number and the average is taken over the Maxwell-Boltzmann distribution. Then we have the form of the reaction rate in (7.1) as

$$\lambda_{12\cdots k} = (\rho N_A)^k \frac{N_1}{N_1! N_2! \cdots N_k!} \langle\sigma v\rangle_{12\cdots k}, \tag{7.2}$$

where N_j is the particle number of element j among the $k + 1$ bodies and $N_j!$ is necessary to avoid multiple counting of reactions with identical particles. For a decay with half-life $\tau_{1/2}$, the rate is written as

$$\lambda_\beta = \frac{\ln 2}{\tau_{1/2}}. \tag{7.3}$$

M. Hashimoto et al., *Big-Bang Nucleosynthesis*, SpringerBriefs in Physics,
https://doi.org/10.1007/978-981-13-2935-7_7

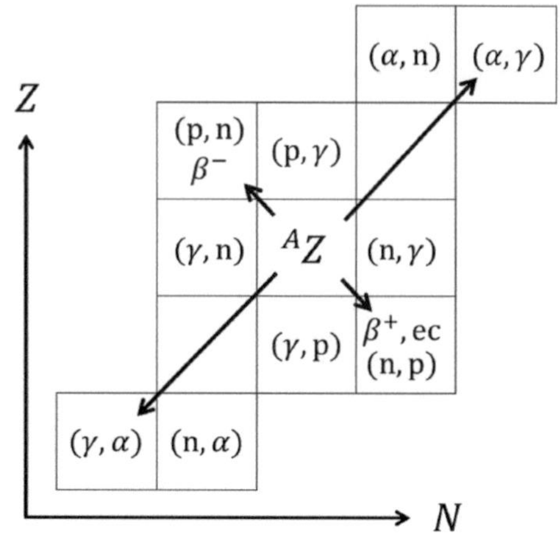

Fig. 7.1 Typical branches of nuclear reactions from a specific nucleus ^AZ on the N-Z plane

Figure 7.1 shows the branches of nuclear reactions relevant for nucleosynthesis on the N-Z plane, where N and Z are the neutron and proton numbers, respectively. This is a section of the chart of nuclides and indicates paths of nuclear reactions with respect to a specific nucleus ^AZ of mass number $A = N + Z$. The paths β^-, β^+, and ec denote β^--decay, β^+-decay, and electron capture, respectively.

The evolution of abundance can be obtained from (7.1) by taking into account the flow of nuclides shown in Fig. 7.1. For example, BBN starts from the production of deuterium through p(n,γ)D. Then it is followed by the destruction via D(p,γ)^{3}He, D(d,n)^{3}He, and so on.

Generally, (7.1) is written as

$$\frac{dY_j}{dt} = f_j\,(Y_1, Y_2, \cdots, Y_n), \qquad (j = 1, 2, \cdots, n). \tag{7.4}$$

We must solve a set of differential equations (7.4) without numerical instability. To this end we adopt the implicit integration method. Let $Y_j^{(i)}$ and $Y_j^{(i-1)}$ be the number fractions of the j-th nucleus at the time steps i and $i-1$, respectively, and then (7.4) reduces to the full implicit backward difference equation.

$$\frac{Y_j^{(i)} - Y_j^{(i-1)}}{\Delta t^{(i)}} = f_j\,(Y_1^{(i)}, Y_2^{(i)}, \cdots, Y_n^{(i)}), \tag{7.5}$$

where $\Delta t^{(i)}$ is the time interval at the i-th time step.

When the right-hand side of (7.5) would be evaluated in terms of the solutions $Y_j^{(i-1)}$ at the previous time step as $f_j\,(Y_1^{(i-1)}, Y_2^{(i-1)}, \cdots, Y_n^{(i-1)})$, it is the explicit method. We will often meet numerical instability, unless we take a infinitesimal time step.

On the contrary, in our implicit method, the right-hand side contains the variables at the i-th time step. We set

$$Y_j^{(i)} = Y_j^{*(i-1)} + \delta Y_j^{*(i-1)}, \tag{7.6}$$

$$Y_j^{*(i-1)} = Y_j^{(i-1)} + \Delta Y_j^{(i-1)}, \tag{7.7}$$

where $Y_j^{*(i-1)}$ is a trial value for $Y_j^{(i)}$. Then (7.5) is written as

$$\frac{\delta Y_j^{*(i-1)} + \Delta Y_j^{(i-1)}}{\Delta t^{(i)}} = f_j(Y_k^{*(i-1)} + \delta Y_k^{*(i-1)}), \quad (k = 1, 2, \cdots, n). \tag{7.8}$$

The right-hand side is expanded to the first-order term as follows:

$$f_j(Y_k^{*(i-1)} + \delta Y_k^{*(i-1)}) = f_j(Y_k^{*(i-1)}) + \sum_{k=1}^{n} \frac{\partial f_j}{\partial Y_k^{*(i-1)}} \delta Y_k^{*(i-1)}.$$

Consequently, we obtain a set of simultaneous linear equations for $\delta Y_j^{*(i-1)}$:

$$\sum_{k=1}^{n} \left(\frac{\delta_{jk}}{\Delta t^{(i)}} - \frac{\partial f_j}{\partial Y_k^{*(i-1)}} \right) \delta Y_k^{*(i-1)} = f_j(Y_k^{*(i-1)}) - \frac{\Delta Y_j^{(i-1)}}{\Delta t^{(i)}}, \quad (j = 1, 2, \cdots, n). \tag{7.9}$$

When we set a guess evaluated from the previous solutions by

$$\Delta Y_j^{(i-1)} = \left(Y_j^{(i-1)} - Y_j^{(i-2)} \right) \frac{\Delta t^i}{\Delta t^{i-1}},$$

then we seek solutions of (7.9) iteratively. The iteration is terminated when the maximum value of the relative corrections $|\delta Y_j^{*(i-1)}/Y_j^{(i)}|$ becomes less than a critical value. We have just obtained a set of solutions $Y_j^{(i)}$ at the i-th time step.

It is noted that the mass fraction is given by

$$X_j = A_j Y_j, \tag{7.10}$$

where A_j is the atomic mass of element j. We check that the normalization

$$\sum_{j=1}^{n} X_j = 1$$

is sustained at every time step.

As an example, we show in Fig. 7.2 the evolution of mass fraction X_j through the hot CNO cycle, which is calculated toward the exhaustion of hydrogen. The network consists of 12 nuclei: ^{1}H, ^{4}He, ^{12,13}C, $^{13-15}$N, $^{14-17}$O, and ^{17}F. The included reactions have only positive Q-values and β-decays. The initial composition is set

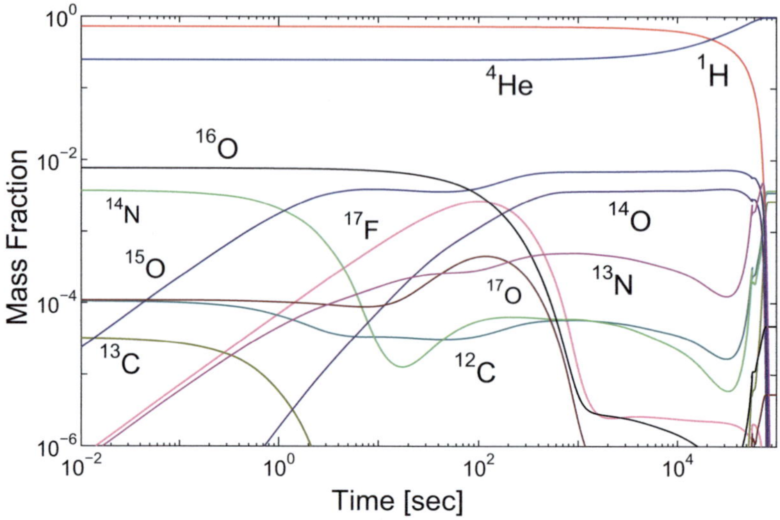

Fig. 7.2 Abundance changes through the hot CNO cycle toward the exhaustion of hydrogen

to be the solar system abundance. Density is $10^2\,\mathrm{g\,cm^{-3}}$. Temperature is around 2×10^8 K and slightly changes during the course of the evolution in accordance with the nuclear energy generation [2]. It is noted that the ordinary CNO cycle operates at $T \simeq 3 \times 10^7$ K in main sequence stars.

References

1. Hashimoto, M., Arai, K.: The nuclear reaction network I. Nucleosynthesis from early universe to supernovae. Phys. Rep. Kumamoto Univ. **7**, 47–63 (1985)
2. Müller, E.: Nuclear-reaction networks and stellar evolution codes – the coupling of composition changes and energy release in explosive nuclear burning. Astron. Astrophys. **162**, 103–108 (1986)

Chapter 8
Appendix D: Some Useful Numbers

Abstract We summarize physical and astronomical constants in cgs units. We also provide the conversion of peculiar units in nuclear physics and astronomy used in this book.

Keywords Physical constants · Astronomical units

Physical constants

Speed of light in vacuum	c	$2.9979 \times 10^{10}\,\mathrm{cm\,s^{-1}}$
Gravitational constant	G	$6.6742 \times 10^{-8}\,\mathrm{dyne\,cm^2\,g^{-2}}$
Planck constant	h	$6.6261 \times 10^{-27}\,\mathrm{erg\,s}$
Reduced Planck constant	$\hbar$	$1.0546 \times 10^{-27}\,\mathrm{erg\,s}$
Avogadro's number	N_{A}	$6.0221 \times 10^{23}\,\mathrm{mole^{-1}}$
Boltzmann constant	k_{B}	$1.3807 \times 10^{-16}\,\mathrm{erg\,K^{-1}}$
Atomic mass unit	M_u	$1.6605 \times 10^{-24}\,\mathrm{g}$
Electron mass	m_e	$9.1094 \times 10^{-28}\,\mathrm{g}$
Proton mass	m_p	$1.6726 \times 10^{-24}\,\mathrm{g}$
Neutron mass	m_n	$1.6749 \times 10^{-24}\,\mathrm{g}$
Electronic charge	e	$4.8032 \times 10^{-10}\,\mathrm{esu}$
Electron rest energy	$m_e c^2$	$511.00\,\mathrm{keV}$
Proton rest energy	$m_p c^2$	$938.27\,\mathrm{MeV}$
Neutron rest energy	$m_n c^2$	$939.57\,\mathrm{MeV}$
Radiation density constant	a_{B}	$7.5658 \times 10^{-15}\,\mathrm{erg\,cm^{-3}\,K^{-4}}$
Fine structure constant	α	$1/137.04$
Planck time	t_{P}	$5.3912 \times 10^{-44}\,\mathrm{s}$
Planck mass	M_{P}	$2.1765 \times 10^{-5}\,\mathrm{g}$

M. Hashimoto et al., *Big-Bang Nucleosynthesis*, SpringerBriefs in Physics,
https://doi.org/10.1007/978-981-13-2935-7_8

Astronomical constants

Solar mass	$M_{\odot}$	1.9891×10^{33} g
Solar luminosity	$L_{\odot}$	$3.8458 \times 10^{33}\,\mathrm{erg\,s^{-1}}$
Hubble constant	H_0	$h \times 100\,\mathrm{km\,s^{-1}\,Mpc^{-1}}$
Hubble time	${H_0}^{-1}$	$9.778\,h^{-1} \times 10^{9}$ yr
Critical density	ρ_{cr}	$1.878\,h^{2} \times 10^{-29}\,\mathrm{g\,cm^{-3}}$

Conversion of units

Arc second	$1''$	=	4.8481×10^{-6} rad
Electron volt	1 eV	=	1.6022×10^{-12} erg
Fermi	1 fm	=	1×10^{-13} cm
Cross section	1 barn	=	$1 \times 10^{-24}\,\mathrm{cm^2}$
Ångström	1 Å	=	1×10^{-8} cm
Year	1 yr	=	3.1558×10^{7} s
Light year	1 lyr	=	9.4607×10^{17} cm
Parsec	1 pc	=	3.0857×10^{18} cm

Index

M. Hashimoto et al., *Big-Bang Nucleosynthesis*, SpringerBriefs in Physics,
https://doi.org/10.1007/978-981-13-2935-7

MIX
Papier aus verantwortungsvollen Quellen
Paper from responsible sources
FSC® C105338

If you have any concerns about our products,
you can contact us on
ProductSafety@springernature.com

In case Publisher is established outside the EU,
the EU authorized representative is:
Springer Nature Customer Service Center GmbH
Europaplatz 3, 69115 Heidelberg, Germany

Printed by Libri Plureos GmbH
in Hamburg, Germany